T0230604

LONG-TERM EFFECTS OF SEWAGE SLUDGE AND FARM SLURRIES APPLICATIONS

Proceedings of a Round-Table seminar organized by the Commission of the European Communities, Directorate-General Science, Research and Development, Environment Research Programme, held in Pisa, Italy, 25–27 September 1984.

LONG-TERM EFFECTS OF SEWAGE SLUDGE AND FARM SLURRIES APPLICATIONS

Edited by

J. H. WILLIAMS

Ministry of Agriculture, Woodthorne, Wolverhampton, UK

G. GUIDI

Laboratoria per la Chimica del Terreno, CNR, Pisa, Italy

and

P. L'HERMITE

Commission of the European Communities,
Directorate-General Science, Research and Development,
Brussels, Belgium

Taylor & Francis
Taylor & Francis Group

LONDON AND NEW YORK

Published by Taylor & Francis
2 Park Square, Milton Park, Abingdon, Oxon, OX14 4RN
270 Madison Ave, New York NY 10016

Transferred to Digital Printing 2007

British Library Cataloguing in Publication Data

Long-term effects of sewage sludge and farm
slurries applications.
1. Sewage sludge as fertilizer—Environmental
aspects   2. Slurry—Environmental aspects
I. Williams, J. H.   II. Guidi, G.   III. L'Hermite,
P.   IV. Environment Research Programme
363.7'384   TD196.S4

ISBN 0-85334-399-3

WITH 91 TABLES AND 82 ILLUSTRATIONS

© ECSC, EEC, EAEC, BRUSSELS AND LUXEMBOURG, 1985

Publication arrangements by Commission of the European Communities, Directorate-
General Information Market and Innovation, Luxembourg

EUR 9731

LEGAL NOTICE

Neither the Commission of the European Communities nor any person acting on behalf of
the Commission is responsible for the use which might be made of the following
information.

All rights reserved. No part of this publication may be reproduced, stored in a retrieval
system, or transmitted in any form or by any means, electronic, mechanical, photocopying,
recording, or otherwise, without the prior written permission of the publisher.

Publisher's Note
The publisher has gone to great lengths to ensure the quality of this reprint
but points out that some imperfections in the original may be apparent

v

Previous seminars organised by the working party under the auspices
of the EEC Concerted Action have been devoted to the evaluation of the
nitrogen, phosphorus and organic matter content of sewage sludges,
almost invariably over the twelve month period of application.
Depending on the type of sludge they have been shown to be useful
sources of nitrogen, phosphorus and organic matter for grass, cereals
and in land restoration in terms of improving the physical properties of
disturbed lands. However, distance from the treatment work severely
limits the radius of operation. Whereas it is of value to farmers
within reach of the works, the impact which it makes nationally on
reducing the annual fertilizer bill is very small.

On grass/arable farms the benefits from sewage sludge can be
complementary to those of animal slurries in terms of providing an
organic manure with an improved balance of nitrogen, phosphorus and
potassium for grass and arable crops.

This was the first seminar at which the experts from both the sewage
sludge and animal manure sides came together to discuss their common
problems and rightly so. Liquid sludges and animal slurries have much
in common though there are differences which must be borne in mind. The
main components of both products are nitrogen and organic matter, the
availability of the nitrogen depending on treatment and composition of
the organic matter.

The present seminar is devoted to the residual and longer term
benefits of sewage sludges and farm slurries. There are papers which
deal with the availability of soil nutrients from sludges and slurries
treated in different ways, on the phosphate balance in soil and the soil
ameliorating properties of these organic amendments. Changes on storage
and mineralisation in soils after treatment with sludges stabilised by
different methods are also discussed. Lastly, but certainly not least,
is the important aspect of evaluating the composition of sludges and
slurries. Variability in slurry composition often makes it desirable to
obtain an 'on the farm' estimate of nutrient value if they are to be
used efficiently without detriment to the environment.

C O N T E N T S

viii

PART II

B. CHANGES ON STORAGE AND MINERALISATION STUDIES IN SOILS AFTER
TREATMENT

A. CUMULATIVE AND RESIDUALS EFFECTS OF SLUDGES AND

FARM SLURRIES

PART I

Use of digested effluents in agriculture

Soil microorganisms and long-term fertility

Comparison of the efficiency of nitrogen in the cattle and pig slurries prepared according to three methods : storage, aeration and anaerobic digestion

Long-term effects of the landspreading of pig and cattle slurries on the accumulation and availability of soil nutrients

Relationships between soil structure and time of landspreading of pig slurry

Results of large-scale field experiments with sewage sludge as an organic fertilizer for arable soils in different regions of the Netherlands

The cumulative and residual effects of sewage sludge nitrogen on crop growth

Long-term effects of farm slurries applications in the Netherlands

Discussion on Part I

USE OF DIGESTED EFFLUENTS IN AGRICULTURE

M. Demuynck, E.-J. Nyns and H. Naveau
Unit of Bioengineering, University of Louvain,
B-1348 Louvain-la-Neuve, Belgium

Summary

Whereas anaerobic digestion has no effect on the quantity of the waste treated, it has well an effect on its quality and consequently on its fertilizer value. Indeed, first of all, 30 to 40 per cent of the organic matter of the waste digested which can be either manure or sludge, are transformed into methane and the remaining organic matter is more stable. If the total nitrogen content of the waste remains more or less the same, the proportion of ammoniacal nitrogen increases (10 to 70 % increasing are reported) and the proportion of organic nitrogen decreases.
The hygienization effect of the anaerobic digestion process is only partial i.e. it allows the destruction of some of the pathogens (bacteria, viruses) under normal running operations. The digested waste is only free of pathogens after treatment under thermophilic conditions.
Concerning the spreading of the digested effluent on land, the effect of the biomethanation treatment is positive since first, the waste is desodorized and secondly, the waste is liquified, homogenized and less sticky and therefore disappears more quickly into the soil which avoids the suppressing of the vegetation and hinders weeds to come out.
Since the disponibility of the nitrogen of the digested effluents is higher than for effluents which are raw or stabilized by other means, yields of crops high demanding in nitrogen will be improved. Ryegrass and maize are typical examples. With the lowering of the C:N ratio of the digested effluent the depressive effect which occurs normally after the application of sewage sludges can be avoided.
From all these considerations, recommendations for the utilization of digested effluents are proposed.

1. INTRODUCTION

If anaerobic digestion was already well developed in the past for the stabilization of sewage sludges, this process becomes at present currently applied for the treatment of agricultural and industrial wastes. Most of the anaerobically digested effluents are not further treated at all and are used in agriculture as fertilizer or soil conditioner.

This is especially the case for part of the sludges produced in the European Community and for the largest part of the agricultural wastes digested. If the farmers know well what is the fertilizer value of their untreated manures or other farm residues, there is at present no precise estimation of the fertilizer value of the digested agricultural wastes. If digested sewage sludges are already often used in agriculture, their potential users may not know exactly what type of fertilizer manure they have.

It may be asked if the anaerobic digestion process has an effect on the organic waste, if yes to which extent and consequently what it involves in practice. Does anaerobic digestion modify the composition of the waste ? Are the anaerobically digested wastes more free of pathogens than the raw ones ? Is their quality improved by the treatment ? Are the availabilities of their nutrients better or worst after digestion ? Can farmers obtain the same crop yields with the digested effluents than with the raw wastes ? Do they have to modify the methods and the times of waste application ? Can they use them on all types of crops ?

Based on a review of the literature and on known experimental results, a study has been realised attempting to answer these questions. This paper gives the essential conclusions of that study.

2. EFFECT OF THE ANAEROBIC DIGESTION ON THE WASTE COMPOSITION

A general consideration to be repeated here is that the anaerobic digestion or biomethanation process affects only the quality of the waste that is treated and that it does not change its quantity at all. On the contrary, sometimes the volumes of the wastes may be increased; this can be the case for semi-solid or solid wastes which are diluted before entering the digester.

The effect of the process on the quality of the waste will depend on several factors as the biodegradability of the waste, the retention time in the digester and the temperature of digestion. A high content in easily biodegradable matters of the waste, a long retention time and a high digestion temperature may lead to important modifications of the nutrient content of the digested effluent. At the contrary, if the waste is not easily degradable and if the digester is not well operating, there can be practically no modifications at all. It is so that the composition of sewage sludges is in general more affected by anaerobic digestion than the agricultural wastes, especially those containing high quantities of lignin, cellulose, those two organic substances being not or partly bioconverted.

In summary, anaerobic digestion has two effects on the nutrient content of the wastes : first, it decreases their organic content and consequently their carbon content and secondly, it transforms part of the organic nitrogen into ammonia nitrogen. The organic matter can be reduced by almost half for sewage sludges (1). For liquid and semi-solid manures, the organic matter reductions reported are of about 40 per cent (2) (3) whereas for manures with bedding, the reduction is less than 20 per cent (4). The remaining organic matter is more stable than originally.

The proportion of ammoniacal nitrogen related to the total nitrogen content increases from about 5 per cent up to as high as 70 per cent. The mineralizations of nitrogen are again more spectacular for sewage sludges and are more important under heated digestion than cold digestion. This mineralized or ammoniacal nitrogen is more readily available to plants than the organic one. This is thus an improvement of the quality of the treated waste. Nevertheless ammonia can be rapidly lost under certain circumstances; precautions for utilization of the digested effluents have so to be taken so as it will be seen further.

A direct consequence of these two effects is that the carbon to nitrogen ratio decreases after anaerobic digestion. For wastes having originally C/N ratios above 15 (raw sewage sludges, manures with bedding, crop

residues), this is thus also an improvement. Indeed, generally after digestion, their C/N ratios decrease to around 10 and so there is no nitrogen
immobilization once the waste is incorporated in the soil.

Concerning the other nutrients i.e. phosphorous, potassium and other
macro nutrients, they appear to remain unaltered by anaerobic digestion.
Nevertheless, since the dry matter content decreases in the digested effluents, their concentrations increase after digestion when expressed to
the dry matter content.

The same observation is valid for the heavy metals and other trace
elements' content. Although, these potentially toxic elements are precipitated during anaerobic digestion, it appears that once in the soil, these
precipitates will oxidize and so have the same behaviour as those of raw
wastes (5). Concerning the degradation of nonionic surfactants that may be
present in sewage sludges, it appears that anaerobic digestion favours the
production of 4-nonyphenol, a persistent toxic substance. Nevertheless,
there is at present no case of intoxication reported. On the other hand,
the process appears to favour the degradation of most of the insecticides
(7). Although this has still to be verified for full-scale operation, this
may be a point in favour of the anaerobic digestion.

3. REDUCTION OF PATHOGENS BY ANAEROBIC DIGESTION

Both sewage sludge and animal wastes may contain a variety of pathogens which present disease hazards to man and food animals or plants.
They may be bacteria, viruses, helminths, fungi or protozoa.

At laboratory scale, mesophilic anaerobic digestion allows a significant reduction of those pathogens. Indeed, the relative reduction is good
for plant pathogens and parasitic cysts, moderate to good for bacteria,
moderate for viruses (12), and poor for helminths ova. This pathogens'
reduction effect is very similar to conventional aerobic digestion. In
practice, especially in completely mixed digesters (12) (14), this effect
is reduced by short-circuiting and by simultaneous drawing off and feeding
of the digester (15) (16). In that case, there is always a risk that the
mixed liquor does not remain inside the reactor during the whole retention
time; infectious organisms may so be discharged and recontaminate the rest
of the effluent. To obtain at full-scale operation the same disinfection
effect as at laboratory-scale, the waste should be treated in batch or
plug-flow digestion systems. An effluent completely free of pathogens can
only be obtained after digestion at thermophilic temperatures i.e. around
55 °C. Even for helminths eggs, there is a complete inhibition of normal
egg development (17).

Although thermophilic anaerobic digestion is seldom employed until
now, it should be recommended in the future to digest at 55 °C the most
infected wastes as abattoir wastes for example.

4. QUALITY IMPROVEMENT OF THE WASTE BY ANAEROBIC DIGESTION

Besides its two effects on waste composition and disease survival,
anaerobic digestion may also improve the quality of the waste. First of
all, it reduces the odour of the waste. Indeed, both for sludge and for
manure, smells during utilization (spreading) cause more complaints than
any other aspect of the operation. These problems can be minimized and
controlled by using digested effluents. It has for example been calculated
that the emission from 1 ha spread with digested slurry is 10 per cent of
the emission from 1 ha spread with untreated slurry (18). The reduction of
smell could justify the fact that cattle prefer grazing on pastures spread

with digested manures (19). Although the odour reduction is not as complete as the one obtained by continuous aeration, the odour reduction by anaerobic digestion is of higher long term value. Indeed for well anaerobically digested manure, odour remains unchanged even after 120 days of storage (19), and the concentrations of malodorous compounds may even be reduced during that time (20). Comparatively, well aerated manure, has been reported to smell again after two to three weeks (19).

As most of the waste treatments, anaerobic digestion reduces the viscosity of the wastes; consequently they become less sticky and more homogeneous (21) (22) (23). Their improved physical state leads to an easier handling and blockages in spreading equipment are avoided. The digested effluents disappear more quickly into the sward and do not form a mat as they dry which smothers new grass. Consequently, there is no or less opportunity for weeds to grow (22). Weed control occurs also in the anaerobic digester itself. Anaerobic digestion appears to have a quite significant fatal effect on weeds as rumex, millet, effect which should be superior to the one of storage or aeration (22). This weed control is nevertheless restricted to the wastes which undergo the treatment and has consequently limitations. It avoids at least the transmission of weeds by manure and other wastes.

5. INFLUENCE OF DIGESTED EFFLUENTS ON THE SOIL

Organic waste may be used as soil conditioner so as to restore derelict and disturbed land e.g. mining spoil, landfill sites, unproductive land... Such soils have very low levels of nutrients and the lack of organic matter can render them more susceptible to physical damage, particularly compaction from heavy machinery resulting in poor root penetration, low infiltration rates with consequent risk of runoff, erosion and pollution (24).

By increasing the carbon content of the soil, organic wastes applications increase aggregation, decrease bulk density, increase water holding capacity and hydraulic conductivity (25). Furthermore organic matter and soil aggregation inversely related to runoff volumes and sediment loss.

Digested organic wastes can be used to soil restoration as well than the organic wastes. Even it has been seen below point 2 that the organic matter content is reduced on average by 30-40 per cent, this is nevertheless largely compensated by the fact that the organic matter of digested effluents is more stable than the organic matter of raw organic wastes and will consequently degrade more slowly in the soil. The effect of digested effluents on soil physical conditions is consequently more permanent than for raw wastes. As for example, the aggregate stability of soil appears to be particularly improved by anaerobic wastes. Guidi et al., (26) have compared the Water Stability Index (WSI) after application of anaerobically digested sludge, aerobically digested sludge and compost of sludge with domestic refuse. The effects on WSI of the different amendments are measured at high loading rates i.e. corresponding to 150 tons/ha of manure on the organic carbon basis which means for the sludge about 30 tons dry solids/ha. These results are reported in Fig. 1. It can be observed that the stabilizing effect of the anaerobic sludge treatment is always higher and longer lasting than that of the other treatments. Pagliai et al., (27) attribute this fact to the relative large amounts of stable organic compounds such as lignin, cellulose, lipids and humic-like materials which are not modified during the anaerobic digestion. These classes of compounds are highly reactive and can interact directly with soil surfaces, thus strenghtening the aggregates. Moreover, such chemical compounds are rather resistant to microbial degradation, and could so explain the lower rate of decay observed

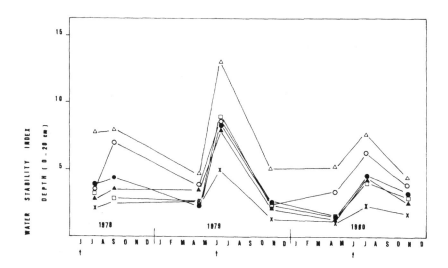

Fig. 1. <u>Effects of treatments on water stability index (WSI). Addition rate corresponds to 150 tons/ha of manure on the organic carbon basis</u>

X: Control; O : Aerobically digested sludge (AS); △ : Anaerobically digested sludge (ANS); ● : Compost of AS + organic fraction of domestic refuse; ▲ : Compost of ANS + organic fraction of domestic refuse, □ : Manure.

in Water Stability Index of the anaerobic sludge treated samples.

Nevertheless, so as to obtain measurable beneficial effects on soil physical conditions, applications rates as high as those used for farmyard manure i.e. about 30 to 50 tons dry solids/ha are required. Therefore, the digested effluents with the highest dry solids content are particularly indicated; these can be dry or dewatered digested sludges or digested manure with bedding or abattoir wastes...

Significant effects on soil improvements cannot be expected from application rates of liquid digested effluents (1-2 tons/ha) which are commonly in use. The small beneficial results may nevertheless accrue from repeated applications, at least if there are no problems of excessive contents of nitrogen or heavy metals in the liquid effluent.

6. <u>AVAILABILITY OF THE NUTRIENTS OF THE DIGESTED EFFLUENTS</u>

The biomethanation process affects essentially the nitrogen availability of the digested wastes but apparently does not influence the availability of the other nutrients. Since during anaerobic digestion part of the organic nitrogen is transformed into ammonia nitrogen and since ammonia-N is more readily available to the plant (assumed to be equal to the N-availability of N-inorganic fertilizer), the global nitrogen availability of digested effluents is higher than that of raw wastes. The total available nitrogen of digested effluents will essentially depend on the proportion of the ammonia nitrogen (in per cent of the total nitrogen) since it is in general assumed that ammoniacal nitrogen and organic nitrogen are respectively 100 and around 20 per cent available (28).

Compared to organic nitrogen, ammoniacal nitrogen presents more risks for volatilization or nitrification and may consequently be lost for the plant. The factors which affect the most the degree of NH_4^+-N losses are the post-treatment of the digested waste and its application conditions. Dewatering or drying are possible post-treatments mostly applied for sewage sludges. Much of the ammonium-N is lost after such treatment. It is so that the amount of nitrogen in liquid digested sludge is up to 3 times that in dried digested sludge (29) and about 20 per cent nitrogen availabilities are reported for such sludges (28). Storage is another possible post - treatment applied for both digested sludges and farm wastes. In such case, ammonia losses are not as high as for dewatering or drying. It appears in general that stored digested effluent has a lower nitrogen availability than freshly digested effluent, but it has nevertheless a higher efficiency than dewatered digested effluent. Although this would need more confirmation, it appears also that the losses are more or less identical for untreated and digested wastes during their storage (30).

Among the application conditions having an effect upon the nitrogen effectiveness of digested effluents, we find :
- the timing of effluents application,
- the soil type, the soil cover, and its environmental conditions i.e., the rainfall and soil temperature conditions,
- the way of application.

Application of digested effluents over autumn and winter period can lead to high losses of nitrogen. Particularly, for autumn-early winter application, the soil may still be warm enough to allow nitrification of the ammoniacal-N and so it is possible that much of the nitrogen may be lost in drainage (29). In heavy textured soils, autumn and winter rainfalls maintain a high degree of moisture in the soil which leads again to ammoniacal-N nitrification with consequent losses. Nevertheless, it has been demonstrated in the United Kingdom, that winter applications of digested sludges do not lead to nitrogen losses, at least in light textured soil (29). Table I. illustrates this particular aspect. By compring the yield response to sludge and fertilizer applied at these times relative to the response to fertilizer N applied in April (this last being equivalent to 100 %), it can be seen that the relative performances of liquid digested sludge are highest from the January and February applications. Grass roots are probably able to absorb ammoniacal nitrogen and this is retained in the surface soil until temperatures are high enough for nitrification and growth to take place.

The influence of the texture of the soil on ammonia losses is well illustrated by Table I. Compared to the light textured soils, the heavy soils remain moister during autumn and winter rainfalls. Under such conditions, the ammonia-N is easily nitrified and is so lost (36). On the other hand, light textured soils are drier in summer conditions and this will favour the ammonia volatilization. The fact that most of the liquid digested effluents have a high water content is nevertheless benefic for summer applications. It has been observed for example that liquid digested sludge applied to drying soil in summer gives a quicker growth than conventional N-fertilizer applied at the same time (28).

Table I. Yields from liquid digested sludges and fertilizers expressed as a percentage of fertilizer N applied in April (30)

	Nov. 23	Jan 10	Feb 28	Apr 1
Clay loam : Sludge (heavy soil)	33	94	99	94
Fertilizer	20	75	73	100
Sandy clay loam : Sludge (light soil)	101	115	101	83
Fertilizer	67	97	74	100

The weather conditions under which the waste is applied influence greatly the ammonia volatilization. It is mostly affected by high air temperature (31) and wind, particularly on bare soils. Consequently, digested effluents should preferably be applied on growing crops covering the soil and under rainy weather conditions.

The application method of the digested effluent is also of great importance for ammonia losses. Injection of the influent appears to be the best method for avoiding such losses, especially compared to surface application but is not very applied in practice. In surface application, losses can nevertheless be minimized by immediate incorporation into the soil following the spreading (32).

7. PLANT RESPONSE TO DIGESTED EFFLUENTS

As far as the response of the plant is considered, anaerobically digested effluents appear to have a good effect. Indeed, since anaerobic digestion produces effluents with lower carbon to nitrogen ratios (C/N = 10), the effect of nitrogen immobilization does not occur after the application of the digested waste whereas it can well occur with some of the untreated wastes (sewage sludges, farm wastes with a high carbon content). Since digested effluents contain a higher percentage of directly available nitrogen i.e. ammonia nitrogen, the time response of the plant is more immediate compared to fresh wastes. It has for example been observed that grass grows faster with digested slurry compared to untreated slurry and can so be harvested 10 days earlier (33).

A restriction has nevertheless been observed concerning the seedling emergence, especially in laboratory experiments. Due to free ammonia, there is a risk that seed germinates poorly if sown on cultivated ground which has just been treated with freshly digested effluents (28,34). Normally, most ammonia is absorbed in soil as the ammonium ion, but at high pH levels, particularly above 7.5 increasing proportions remain as free ammonia from which levels in excess of 2.5 mg/l are toxic to many plants (28). This effect dissappears within one week and has until now not been reported in field experiments.

Concerning the crop yields that can be observed, it is generally accepted that liquid digested wastes give increases in yield of crops similar to those produced by equivalent amounts of nitrogenous fertilizer (35,22). Crops high demanding in nitrogen so as ryegrass and maize are particularly indicated for the utilization of such wastes. Cereals would

appear to profit less than other crops and there may be even an insuffi-
ciency in available nitrogen to carry the crop to maturity (28). In a 6
years field trial on ryegrass comparing digested manure to untreated and
aerated manure, Besson (22) observes that grass yields obtained are more or
less identical for the 3 differently treated manures. In fact, grass yields
appear to be more function of the type of manure (bovine cattle or pig) and
of the application rates. The feed value of the crop (grass or maize) expressed in digestible crude albumin appears to be unchanged for all the
treatments (29). Both for the utilization of digested sludge or manure on
mixed herbage, it has nevertheless been observed that clover tends to be
suppressed, but this effect is less than that of ammonium nitrate (22,28).

8. CONCLUSIONS

Anaerobic digestion changes only the quality of the waste i.e. compared to raw waste, digested effluents have a lower content of organic
matter, a higher content of ammonia nitrogen, are more easily usable and
are more disinfected although not completely.

Dried digested effluents or solid digested wastes should preferably
be used for restoration of unfertile soils so as to improve their physical
conditions. Concerning the recommendations that may be given for the utilization of liquid digested effluents, the essential point is to take the
maximum advantage of the ammonia nitrogen that contain digested waste and
therefore to avoid ammonia losses. Therefore the digested effluents should
never be spread under dry and sunny weather. In case of use on bare soils,
surface application should immediately be followed by the ploughing in of
the waste. An ideal cultural practice is for example to combine the incorporation of liquid digested effluents with that of crop residues rich in
carbon e.g. stubble.

Defining the optimal times for digested effluents utilization is
rather difficult and depends in fact on several factors as climates, types
of crops, types of soils... On a general rule, the wastes should be applied
when crops need nutrients for growing, otherwise, there is always a risk
for nitrogen leaching especially on soils having a low organic matter content. Winter applications can be allowed only on light textured and levell-
ed soils and when temperatures are low. In any case, winter spreading on
slopes should be forbidden because there may be risks for superficial
runoff.

The typical way for the utilization of digested effluents on grass is
applying them firstly before the growth of the new shoots and than after
each cut. Indeed spreading on new grass is dangerous and can lead to the
necrosis of the new shoots and well developed grass may be easily burned
when too high a quantity of ammonia is applied in once. Concerning maize
growth, the best season of application appears to be a first time before
sowing and a second time, somewhat 2 months after (22). For cereals although it is not recommended, the best yields appear to be obtained with
January and February applications.

This study has been supported by grant ECI-1012-B7210-83-B from the
Commission of the European Communities.

REFERENCES

1. HALL, J.E., 1983. Predicting the nitrogen values of sewage sludges. In Characterization, Treatment and Use of Sewage Sludge, Proceedings of the third European Symposium, Brighton, 26-30 September 1983. Commission of the European Communities, L'Hermite P. and Ott H. eds., Reidel D., Publishing Company, Dordrecht, Holland, in the press.

2. DEMUYNCK, M., NYNS, E.-J., 1984. Biogas plants in Europe - practical handbook. In Energy from Biomass, Série E, Vol. 6., Grassi G. and Palz, W., eds., D. Reidel Publishing Company, Dordrecht, Holland, in the press.

3. BESSON, J.-M., LEHMANN, V., ROULET, M., WELLINGER, A., 1982. Influence de la méthanisation sur la composition des lisiers, Revue Suisse Agric. 14 (3), pp. 143-151.

4. CHAUSSOD, R., SANCHEZ, C., DUMET, M.C., CATROUX, G., 1983. Influence d'une digestion anaérobie (méthanisation) préalable sur l'évolution dans le sol du carbone et de l'azote des déchets organiques. D.G.R.S.T. - Action concertée "Valorisation énergétique des déchets agricoles" - Aide n° 78-7-2909. Rapport de l'I.N.R.A. - Laboratoire de Microbiologie des sols - BV 1540-21034 Dijon - Cedex.

5. BALBWIN, A., BROWN, T.A., BECKETT, P.H.T., ELLIOTT, G.E.P., 1983. The forms of combination of Cu and Zn in digested sewage sludge. Water. Res., 17 (12), pp. 1935-1944.

6. SCHAFFNER, C., BRUNNER, P.H. and GIGER, W., 1983. 4-NONYLPHENOL, a highly concentrated degradation product of nonionic surfactants in sewage sludge. In Characterization, Treatment and Use of Sewage Sludge, Proceedings 3d European Symposium, Brighton, 26-30 September 1983. Commission of the European Communities, L'Hermite P. and Ott H. eds, Reidel D. Publishing Company, Dordrecht, Holland, in the press.

7. LESTER, J.N., 1983. Presence of organic micropollutants in sewage sludges, in Characterization, Treatment and Use of Sewage Sludge, Proceedings third European Symposium, Brighton, 26-30 September 1983. Commission of the European Communities, L'Hermite, P. and Ott, H. eds., Reidel D. Publishing Company, Dordrecht, Holland, in the press.

8. TURNER, J., STAFFORD, D.A., HUGHES, D.E., CLARKSON, J., 1983. The reduction of three plant pathogens (Fusarium, Corynebacterium and Globodera) in anaerobic digesters. Agricultural Wastes, 6, pp. 1-11.

9. CARRINGTON, E.G., 1980. The fate of pathogenic micro-organisms during wastewater treatment and disposal. Technical report TR 128, Stevenage Laboratory, Water Research Centre, 58 pp.

10. PIKE, E.B., 1980. The control of Salmonellosis in the use of sewage sludge on agricultural land. In Characterization, Treatment and Use of Sewage Sludge, Proceedings of the 2nd European Symposium, Vienna, October 21-23, 1980; P. L'Hermite and H. Ott eds, D. Reidel Publishing Company, Dordrecht, Holland, pp. 315-327.

11. WILLINGER, H., THIEMANN, G., 1983. Survival of resident and artici-
fially added bacteria in slurries to be digested anaerobically. In
Proceedings of a joint workshop of expert groups of the Commission of
the European Communities, German Veterinary Medical Society (DVG) and
Food and Agricultural Organisation; D. Strauch ed., Institute for Animal Medicine and Animal Hygiene, University of Hohenheim, Stuttgart,
Federal Republic of Germany, pp. 210-217.

12. LUND, E., LYDHOLM, B. and NIELSEN, A.L., 1982. The fate of viruses
during sludge stabilization, especially during thermophilic digestion.
In Disinfection of Sewage Sludge : technical, economic and microbiolo-
gical aspects, Proceedings of a Workshop, Zurich, May 11-13, 1982. A.M.
Bruce, A.H. Havelaar, P. L'Hermite, eds., D. Reidel Publishing Company,
Dordrecht, Holland, pp. 115-124.

13. CARRINGTON, E.G., HARMAN, S.A., 1983. The effect of anaerobic digestion
temperature and retention period on the survival of Salmonella and
Ascaris ova, presented at WRC Conference on stabilisation and disinfection of sewage sludge, Paper 19, Librarian, WRC Processes, Stevenage,
13 pp.

14. BLACK, M.I., SCARPINO, P.V., O'DONNELL, C.J., MEYER, K.B., JONES, J.V.,
KANESHIRO, E.S., 1982. Survival rates of parasite eggs in sludge during
aerobic and anaerobic digestion. Applied and Environmental Microbiology, 44 (5), pp. 1138-1143.

15. GOLUEKE, C.G., 1983. Epidemiological aspects of sludge handling and
management. Biocycle, 24, (3), 52-58.

16. MUNCH, B., SCHLUNDT, J., 1983. The reduction of pathogenic and indicator bacteria in animal slurry and sewage sludge subjected to anaerobic
digestion or chemical disinfection. In Proceedings of a joint workshop
of expert groups of the Commission of the European Communities, German
Veterinary Medical Society (DVG) and Food and Agricultural Organization; D. Strauch ed., Institute for Animal Medicine and Animal Hygiene,
University of Hohenheim, Stuttgart, Federal Republic of Germany, pp.
130-149.

17. KIFF, R.J., LEWIS-JONES, R., 1983. Factors that govern the survival of
selected parasites in sewage sludges. Pollution Research Unit Umist,
Manchester.

18. KLARENBEEK, J.V., 1982. Odour measurements in Dutch agriculture :
current results and techniques. Research Report 82-2, Institute of
Agricultural Engineering, Wageningen, The Netherlands.

19. WELLINGER, A., 1983. Anaerobic digestion - the number one in manure
treatment with respect to energy cost ? In Proceedings of a joint workshop of expert groups of the Commission of the European Communities,
German Veterinary Medical Society (DVG) and Food and Agricultural Organization D. Strauch, ed., Institute for Animal Medicine and Animal
Hygiene, University of Hohenheim, Stuttgart, Federal Republic of
Germany, pp. 163-183.

20. VAN VELSEN, A.F.M., 1981. Anaerobic digestion of piggery waste. Ph. D.
Thesis of the Agric. University, Wageningen, 103 p.

21. Anon., 1983. Practical guidelines for the farmer in the EC with respect to utilization of animal manures.

22. BESSON, J.M., 1983. Personal Communication.

23. PIKE, E.B., DAVIS, R.D., 1983. Stabilisation and disinfection - their relevance to agricultural utilisation of sludge, presented at WRC Conference on stabilisation and disinfection of sewage sludge, Paper 3, Session 1, Librarian, WRC Processes, Stevenage, 30 pp.

24. HALL, J.E., VIGERUST, E., 1983. The use of sewage sludge in restoring disturbed and derelict land to agriculture. In Utilization of Sewage Sludge on Land : Rates of Application and Long-term Effects of Metals, Proceedings of a Seminar, Uppsala, June 7-9, S. Berglund, R.D. Davis and P. L'Hermite eds., D. Reidel Publishing Co., Dordrecht, Holland pp. 91-102.

25. KHALLEL, R., REDDY, K.R., OVERCASH, M.R., 1981. Changes in soil physical properties due to organic waste applications : a review Journ. of Environmental Quality, 10, pp. 133-141.

26. GUIDI, G., PAGLIAI, M., GIACHETTI, M. 1981. Modifications of some physical and chemical soil properties following sludge and compost applications. In The Influence of Sewage Sludge Application on Physical and Biological Properties of soils, Proceedings of a seminar organized jointly by the Commission of the European Communities, Directorate - General for Science, Research and Development and the Bayerische Landesanstalt für Bodenkultur und pflanzenbau, Munich, Federal Republic of Germany, held in Munich, June 23-24, pp. 122-130.

27. PAGLIAI, M., GUIDI, G., LAMARCA, M., GIACHETTI, M., LUCHAMANTE, G., 1981. Effects of sewage sludges and composts on soil porosity and aggregation. J. Environ. Qual. 10, pp. 556-561.

28. COKER, E.G., 1978. The utilization of liquid digested sludge. Paper 7 in WRC Conference, "Utilization of Sewage Sludge on Land". Water Research Centre, 1979 (Stevenage, Herts, SG1 1TH).

29. TOUSSAINT, B., MARCIN, C., 1981. Comparaison de lisiers de bovins avant et après méthanisation. Internal report of the "Laboratoire d'Ecologie des Prairies", Prof. J. Lambert, UCL, B-6654 Michamps (Longville).

30. HALL, J.E., WILLIAMS, J.H., 1983. The use of sewage sludge on arable and grassland. In Utilisation of Sewage Sludge on Land Rates of Application and Long-term Effects of Metals, Proceedings of a seminar, Uppsala, June, 7-9, S. Berglund, R.D. Davis and P. L'Hermite eds., D. Reidel Publishing Co., Dordrecht, Holland, pp. 22-35.

31. BEAUCHAMP, E.G., KIDD, G.E., THURTELL, G., 1978. Ammonia volatilization from sewage sludge applied in the field. J. Environ. Qual., 7, pp. 141-146.

32. RYAN, J.A., KEENEY, D.R., WALSH, L.M., 1973. Nitrogen transformations and availability of an anaerobically digested sewage sludge in soil. J. Environ. Quality. 2, pp. 489-492.

33. STEPHAN, B. 1983. Personal communication.

34. GUIDI, G., HALL, J.E., 1983. Effects of sewage sludge on the physical and chemical properties of soils, in Characterization, Treatment and Use of Sewage Sludge, Proceedings, 3d European Symposium, Brighton, 26-30 September 1983. Commission of the European Communities, L'Hermite P. and Ott H. eds., Reidel D. Publishing company, Dordrecht, Holland, in the press.

35. WILLIAMS, J.H., 1979. Utilisation of sewage sludge and other organic manures on agricultural land. In Treatment and Use of Sewage Sludge, Proceedings of the 1st European Symposium, Cadarache, 13-15 February 1979, D. Alexandre & H. Ott eds, Commission of the European Communities, pp. 227-242.

36. CHAUSSOD, R., 1980. Valeur fertilisante des boues résiduaires. In Characterization, Treatment and Use of Sewage Sludge. Proceedings of the 2nd European Symposium, Vienne, October 21-23, 1980; P. L'Hermite and H. Ott, eds., D. Reidel Publishing Company, pp. 448-465.

SOIL MICROORGANISMS AND LONG-TERM FERTILITY

U. TOMATI, A. GRAPPELLI, E. GALLI

Istituto di Radiobiochimica ed Ecofisiologia Vegetali - Consiglio Nazionale delle Ricerche - Area della Ricerca di Roma; Via Salaria Km 29.300 00016 Monterotondo Scalo (Roma)

Summary

Sludge supplies to soil many plant nutrients, especially N and P, and organic matter, which stimulates microbial activities from which soil biological cycles depend. The biological cycles are responsible of the mineralization and the biosynthesis of many active metabolites.

Four years experiments were carried out on maize. Heavy doses of both aerobic and anaerobic sludge were supplied. Microbial population, oxygen consumption, auxin production, total nitrogen and nitrate content were followed every year in the soil, before treatments. The same parameters in soil, nitrate content and nitrate reductase activity in plant were followed every year at the emergence time. Crop production as q /ha and protein/ha was assayed every year too.

Sludge supply stimulated soil oxygen consumption, which may be considered as an index of increased microbial population and its activities. As a consequence of more efficient mineralization, high content of available ions, in particular $N-NO_3^-$, has been recorded. Phytohormone production by soil microorganisms is strongly stimulated too. Crop yield and quality improved especially after aerobic treatment.

1. INTRODUCTION

"Soil fertility" is a frequently used term for indicating the "inherent capacity of soil to supply nutrients to plants in adequate amounts and in suitable proportions" (2). This is a wide definition which imply that soil fertility is dependent both on chemical, physical, mechanical factors and on biological ones, like microflora and its related activities. These last factors have a great effect on soil nutrient status affecting the nutrient supply to the plant via mineralization and influencing nutrient uptake via biosynthetic activities. Agronomically, the term "soil fertility" is mainly related to crop production. Soil fertility increases when crop production and quality improve.

In the last years, sludge is more and more supplied to crop land, because, aside from all kinds of questions related to public health, it usually benefits agriculture acting as a soil conditioner and as a source of many macro and micro plant nutrients. Soil and crop improvement after sludge supply has been studied in many different ways in order to ascertain the effective fertilizing value of sludge (4, 5, 6, 7).

In spite of this, many questions, some of them regarding the point if sludge should be considered as fertilizers or wastes, are still object of discussion. Sludge especially benefits morfologically or drastically disturbed lands, the same effects being less evident in crop land. In this way, several non arable lands could be recovered to agriculture and crop land could be improved too. The more reliable term for establish if soil fertility has been enhanced by treatments is crop production, although study performed on soil, plant and crop characteristics could help to explain the why.

When sludge is supplied to soils, two kinds of effects could be considered:
- the short-term effect, mainly related to directly available plant nutrients and to rapidly mineralizable materials;
- the long-term effect, mainly related to the transformation of the remander and to its effect on soil properties.

Some important questions are connected to these topics, especially to the second one, which is the theme of this meeting:
1) Sludge is often supplied to crop land on the basis of total nitrogen content. Only about 50% is mineralized in the first year. What is the fate of the remander in the long run and when repeated heavy amounts are supplied?
2) Sludge is more organic than mineral. What is the fate of organic matter and the ways by which humus is produced?
3) Sludge improves the physical properties of soil. Is that only a physical-chemical process or are microorganisms also involved in increasing soil aggregates, e.g. via polisaccarides?
4) Sludge supply could modify biological equilibrium in soil. Can biological cycles give informations about the health status of soil and its fertility after sludge treatment?
5) Soil microorganisms not only mineralize organic matter, but are responsible of several processes of biosynthesis. In consequence, different kinds of metabolic products are excreted into soil. What is the role of microbial metabolites in crop improvement?

The present paper mainly deals with the last question.

2. EXPERIMENTAL DETAILS

With the aim to determine the effect of repeated heavy doses of sludge on crop, four years experiments were conducted in open field with maize as test plant, in plots treated with inorganic fertilizer, as a control, and urban sewage sludge. Sludge or inorganic fertilizer were supplied every year before sowing. Before fertilizer supply, microflora, oxygen consumption, auxin production, nitrate and total nitrogen content in soil were tested. The same parameters were tested during the vegetative cycle of the culture (emergence). At the same time, nitrate reductase, as index of nitrate utilization, and nitrate content in the plant were tested. Crop yield has been determined as q /ha and as protein/ha.

Soil Composition:

pH	Clay	Silt	Sand	C.E.C.	O.M.	N_{tot}	P_2O_5	K_2O
5.8	10%	14%	76%	13,4%	0.9%	0.075%	trace	1.31%

Size of Plots: 500 m²

Sludge: from municipal plants

Sludge composition:

(% d.w.)	C	N		P_2O_5			K_2O		D.M.
	tot	tot	inorg	tot	inorg	ass	tot	ass	
Aerobic Sludge	27.54	4.48	0.15	4.29	3.93	1.57	0.69	0.57	1.6
Anaerobic Sludge	27.99	2.84	0.36	1.17	1.08	0.21	0.13	0.06	3.5

Sludge distribution: spreading

Treatments:

Mineral (control)

(Kg/ha)	N	P_2O_5	K_2O
	250	120	120

Aerobic Sludge   2280 m³/ha equivalent to 200 q O.M. and 1518 Kg N

Anaerobic Sludge   900 m³/ha equivalent to 204 q O.M. and 1101 Kg N

Culture: maize F_1 DF 38.

3. RESULTS

Results are reported in percentage as compared to mineral fertilizer treatment, in order to make evident the possible improvement dependent on sludge supply. Data were collected every year into soil before the treatment, in order to have a general view of soil status. Furthermore, the same parameters in rhizosphere soil at the emergence time were collected too. Contemporarly, some metabolic parameters were followed in the plant in order to have a general view of the conditions created in the plant-soil system by sludge treatment.

Microflora and soil activities

Microflora and soil activities were determined every year a) before the treatment and b) at the emergence time.
a) No significant differences were determined for fungi and actinomycetes whereas bacteria profited from sludge treatment only after the third year. This is made evident at the beginning of the fourth year when plots had been treated for three times already. The ratio bacteria/fungi increased. The improvement of a such ratio could be judged favourably as index of soil fertility (1). Soil activities showed a similar behaviour.
b) Microflora, especially bacteria, was strongly enhanced since the first year. Soil activities showed an increase in oxygen consumption, which may be considered an index of organic matter mineralization, and in auxin production, which may be assumed as an index of microbial biosynthetic processes. The data, collected in the rhizosphere, showed that microbial activities create favourable edaphyc conditions for the plant as $N-NO_3^-$ content confirms, and produce hormonal factors which stimulate nutrient uptake and utilization (3, 8,10).

Plant analysis

Plant analysis seemed to support the better edaphyc and environmental conditions for plant growth created by sludge supply. Nitrate content in the plant increased, showing a stimulation in nitrate uptake. As nitrate reductase activity value confirms, the plant is able to control $N-NO_3^-$ uptake and its organization.

Crop yield

There are not remarkable differences between yield (as q/ha and q of protein/ha) obtained after sludge or mineral treatment. However, after aerobic sludge treatment the same slightly improvement is recorded since the first year. After anaerobic sludge treatment, the improvement became evident after the third year, when crop production reached the value of the control.

CROP YIELD

years	1		2		3		4	
	q/ha	protein (q/ha)	q/ha	protein (q/ha)	q/ha	protein (q/ha)	q/ha	protein (q/ha)
Mineral	100	100	100	100	100	100	100	100
Aerobic S.	103	113	113	104	113	119	102	111
Anaerobic S.	92	86	104	86	114	119	105	106

4. CONCLUSIONS

Sludge treatment improves soil fertility as appears from the data regarding crop yield and soil properties. Aerobic sludge are more effective than anaerobic one.

Sludge was supplied without any minerals,so. all nutrients derived either by directly available nutrient content of sludge or they become available after mineralization of more complex materials.

The continue release of nutrients is able to assure sufficient edaphyc condition for plant growth.

Repeated heavy doses of sludge increased the content of organic nitrogen enhancing, in this way, the future fertility of the soil.

Biological cycles were stimulated by sludge supply (9, 11, 12, 13,14) during the vegetative cycle of the culture. As a consequence of the stimulation of the Nitrogen cycle (9, 11, 12, 13, 14) amounts of NO_3^- comparable to those from usual fertilizer, were released into soil.

The mineralization and microbial processes of biosynthesis occur all together.

As a consequence, many biologically active substances are excrete into soil.

Microbial processes (mineralization and biosynthesis) are particularly enhanced in the rhizosphere, the zone where all relationships between plant and soil are strictly dependent and control nutrient uptake and utilization.

In our previous paper (9, 11,12, 13,14) we made evident that growth regulator biosynthesis is strongly enhanced in the rhizosphere during the vegetative cycle of the culture.

The data presented here confirm that microbial active metabolite content into soil increases in the long run.

Numerous papers have stressed the point that microbes and biologically active substances improve the mineral nutrition and enhance the growth, not only by increasing the plant yield, but also confering to the plant better nutritional qualities.

Growth regulators naturally occur into soil and increase after fertilization particularly when organic materials, such as sludge, are supplied.

The positive action of sludge and organic fertilizers on crops cannot only be explained by the content of the mineral nutrients there in, but it is also related to the presence of microorganisms and their metabolites, which increase the uptake and the utilization of inorganic and organic compounds.

Therefore, it is apparent that mediators exist between nutrient sources of the soil and the root system.

Such mediators are the microbes and their secondary metabolites, such as phytohormones, which are nowdays considered as a very interesting and worthy field of study, not only for microbiologist, but also for agronomist.

19

ACKNOWLEDGE

The Autors wish to thank Mr.G. Palma for its technical assistence.

REFERENCES

(1) AHRENS, E. : Beitrag zur Frage der Indikatorfunktion der Bodenmikroorganismen am Beispiel von drei verschiedenen Nutzungsstufen eines Sand-
bodens. Soil Biol. Biochem. 9 , 185-191 (1977).
(2) BRADY, N.C.: The Nature and Properties of Soils. Macmillan Co., New
York, 639p. (1974).
(3) BROWN, M.E. : Plant growth substance produced by microorganisms of soil
and rhizosphere. J. Appl. Bacteriol. 43,443-451 (1972).
(4) COMMISSION OF THE EUROPEAN COMMUNITIES : First European Symposium
Treatment and use of sewage sludge - Cadarache (1979).
(5) COMMISSION OF THE EUROPEAN COMMUNITIES : Characterization Treatment
and use of sewage sludge - Vienna (1980).
(6) COMMISSION OF THE EUROPEAN COMMUNITIES : Concerted Action Treatment and
use of sewage sludge final report - Cost 68 Bis (1981).
(7) COMMISSION OF THE EUROPEAN COMMUNITIES : The Influence of sewage sludge
application on physical and biological properties of soils - Munich
(1981).
(8) GRAPPELLI A. and ROSSI W. : The effect of phytohormones produced by
Arthrobacter sp. on the phosphatase activity in plant roots - Folia
Microbiol. 26, 137-141 (1981).
(9) GRAPPELLI A. and TOMATI U. : Effetti sulla rizosfera, sulla microflora
e sul biochimismo delle piante - Utilizzazione di fanghi in agricoltura
Risultati delle ricerche condotte dal gruppo di lavoro in collaborazio-
ne con la Regione Toscana - Collana P.F. Promozione della qualità del-
l'ambiente AR/2/20-27. Roma (1981).
(10)LIN W., OKON Y. and HARDY R.W. : Enhanced mineral uptake by Zea mays
and Sorghum bicolor roots inoculated with Azospirillum brasiliense.
Appl. Environ. Microbiol. 45, 1775-79 (1983).
(11)TOMATI U., GRAPPELLI A. and GALLI E. : (see ref. 4) pp. 330-336.
(12)TOMATI U., GRAPPELLI A. and GALLI E.: (see ref. 5) pp. 553-561.
(13)TOMATI U. : (see ref.6) pp. 357-371.
(14)TOMATI U., GRAPPELLI A. and GALLI E. : (see ref.7) pp. 229-242.

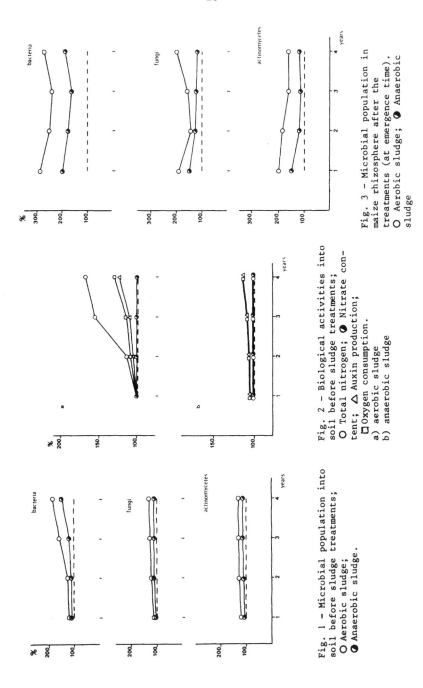

Fig. 1 - Microbial population into
soil before sludge treatments;
O Aerobic sludge;
● Anaerobic sludge.

Fig. 2 - Biological activities into
soil before sludge treatments;
O Total nitrogen; ● Nitrate content; △ Auxin production;
▢ Oxygen consumption.
a) aerobic sludge
b) anaerobic sludge

Fig. 3 - Microbial population in
maize rhizosphere after the
treatments (at emergence time).
O Aerobic sludge; ● Anaerobic
sludge

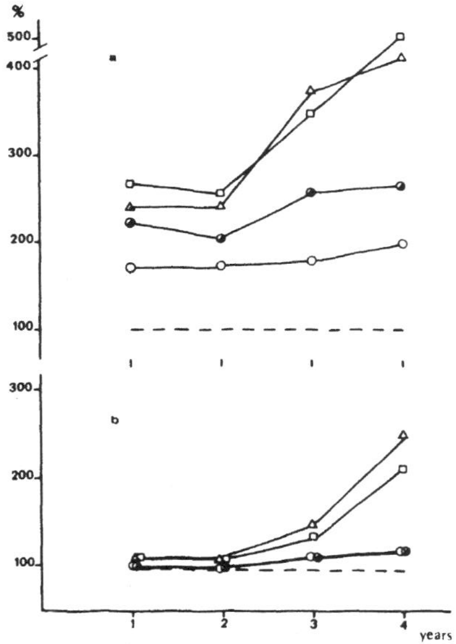

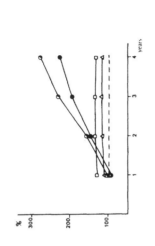

Fig. 5 - Biological parameters in the plant after
sludge treatments (at emergence time).
○ Nitrate content - aerobic sludge;
● Nitrate content - anaerobic sludge;
□ Nitrate reductase activity - aerobic sludge;
△ Nitrate reductase activity - anaerobic sludge.

Fig. 4 - Biological activities in maize rhizosphere
after the treatments (at emergence time).
○ Total nitrogen;
● Nitrate content;
△ Auxin production;
□ Oxygen consumption.

a) aerobic sludge; b) anaerobic sludge.

COMPARISON OF THE EFFICIENCY OF NITROGEN IN THE CATTLE AND PIG SLURRIES
PREPARED ACCORDING TO THREE METHODS: STORAGE, AERATION AND ANAEROBIC
DIGESTION

J.-M. BESSON, R. DANIEL and P. LISCHER
Swiss Federal Research Station for Agricultural Chemistry
and Hygiene of Environment

Summary

This article discusses a) the influence of storage, aerobic and
anaerobic digestion of cattle and pig slurries, and b) the efficiency of nitrogen in the prepared slurries.
    a) The aereted slurries have the lowest mineral nitrogen con-
tent (hence the highest organic nitrogen content), the stocked
slurries are in intermediary position, whilst the anaerobic digested slurries have the highest mineral nitrogen content.
    b) The nitrogen efficiency in the slurries as demonstrated
by pot trials with Italian Rye-grass (Lolium italicum) harvested
thrice yearly, was lower for aerated slurries as compared to the
others; and this not only for the first harvest (direct effect)
but also for the remainder (after-effect, here deceptive). We must
therefore deduce that the organic nitrogen of aerated slurries undergoes the same fate as that of the other slurries: " enter
the great pool of soil organic nitrogen, from which the mineralization rate rises usually 1-3% yearly" (2), and is also only
slightly available by the plants. Lastly, depending on the method
of slurry preparation, the relationship between the fraction of
ammoniacal nitrogen in the slurries and the observed parameters
on the Rye-grass was not obvious. The types and the amounts of
slurries predominate over the effect of the slurry preparation
on the plant behaviour, especially for the pig slurries.

1. INTRODUCTION

    Different methods of treatment of the slurries have been praised
during the last 10 years: beside the common storage, more often than
not joined to the brew of slurries, these are also prepared by aeration
or methanic digestion. These methods are well known (3, 8).
    In addition to a sensitive reduction of the emanations of smells
during the brew, but above all during the spreading of slurries, these
treatments, compared to the storage, have a not negligible influence on
the composition of the treated slurries. Most of the results on this subject have been previously published (6); all of them are concentrated in
the first part of this study. They especially show that the aerated
slurries have the lowest mineral nitrogen content (hence the highest or-

ganic nitrogen content) , the stocked slurries are in intermediary position, whilst the anaerobic digested slurries have the highest mineral
nitrogen content.

This fact has allowed to build up a working hypothesis on which is
established the comparative study of the efficiency of nitrogen in the
slurries prepared according to different treatments: On the one hand, by
the application of the aerated slurries a better effect of nitrogen is
expected at the beginning of the period of vegetation (direct effect of
the mineral nitrogen on the first harvest), but a greater effect after
application of the stocked slurries, and, a fortiori, of the methanic
digested slurries. On the other hand, an upper effect of the aerated
slurries is provided for later in the period of vegetation, according
to the mineralization of the organic nitrogen of these slurries in the
soil (after-effect for the following harvests), and the opposite for
the other slurries.

In order to test this hypothesis, pot experiments were carried out
with Italian Rye-grass during three years. The results of these experiments constitute the second part of this study.

2. EXPERIMENTATION

2.1 Treatment of slurries

Both types of cattle and pig full slurries, which were submitted to
the three treatments storage, aeration and anaerobic digestion, and the
experimental plant for slurry preparation have been previously described
(6). Since, four further preparations have been executed; therefore, the
total number of preparations for the cattle slurries is: 16 times storage, 16 times aeration and 20 times methanic digestion; for the pig
slurries, 18 times for each treatment.

2.2 Pot experiments

The pot experiments were carried out in the years 1979, 1980 and
1981. The treated slurries were applied on the Italian Rye-grass (Lolium
italicum) a few days before the sowing, at a rate of 1.5 g/pot (low
amount or amount 1) or 3.0 g/pot (high amount or amount 2) of total nitrogen contained in the slurries. The quantities of fertilizing matter
brought by the slurries are given in Table I. Further, a mineral fertilization was applied every year, also before the sowing: 2.0 g/pot P_2O_5
(superphosphate) and 6.0 g/pot K_2O (potash 60 %). During the period of
vegetation, after the first and the second harvest respectively, a potash fertilization (1.5 g/pot) was still added. Two experiment soils were
utilized, the one with a neutral reaction (pH=7.0), the other acid
(pH=5.4); their main caracteristics are shown in Table II. Three harvests of Italian Rye-grass were realized every year, a fourth was no
more possible because the growth was practically stopped after the third.
A detailled description of these pot trials was produced by another source (7).

Table I. Quantities of fertilizing matter in the spread slurries

Year	Process	Type of slurries Treatment	Amount ml/pot	N_T	N-NH$_4$	P$_2$O$_5$	K$_2$O
		C a t t l e					
1979	1	storage	765	1.50	0.71	0.91	1.77
	2	aeration	848	1.50	0.53	0.84	1.92
	3	anaer. digest.	769	1.50	0.70	0.81	1.68
		P i g					
	4	storage	267	1.50	1.03	0.94	0.84
	5	aeration	270	1.50	1.13	0.73	0.82
	6	anaer. digest.	254	1.50	1.21	0.87	0.79
	7-12	the same process as 1-6, but the double amount	...	3.00	....	....	....
		C a t t l e					
1980	1	storage	962	1.50	0.63	0.86	1.67
	2	aeration	968	1.50	0.55	0.79	1.52
	3	anaer. digest.	987	1.50	0.71	0.88	1.73
		P i g					
	4	storage	405	1.50	1.11	1.02	0.91
	5	aeration	344	1.50	0.94	0.88	0.79
	6	anaer. digest.	437	1.50	1.14	0.85	0.87
	7-12	the same process as 1-6, but the double amount	...	3.00	....	....	....
		C a t t l e					
1981	1	storage	655	1.50	0.75	0.77	1.59
	2	aeration	773	1.50	0.63	0.94	1.95
	3	anaer. digest.	507	1.50	0.94	0.79	1.13
		P i g					
	4	storage	315	1.50	1.04	0.86	0.62
	5	aeration	342	1.50	1.05	0.84	0.73
	6	anaer. digest.	299	1.50	1.16	0.74	0.67
	7-12	the same process as 1-6, but the double amount	...	3.00	....	....	....

Table II. Properties of the experimental soils

Origin		Liebefeld	Steinhof
pH		7,0	5,4
Organic matter	%	3,1	2,8
P$_2$O$_5$, index		6,3	3,0
K$_2$O, index (mg/100 g)		5,0	1,6
CaCO$_3$		0	0
Clay < 2 µm	%	24	17
Silt 2-63 µm	%	52	22
Sand 63-2000 µm	%	24	61

3. RESULTS AND DISCUSSION

3.1 Influence of the preparation of slurries on their composition

The results concerning the comparison of the slurry composition before and after the preparation are given in Table III. In the main, they
confirm what had been previously published (6).

The influence of the preparations on the pH is significant for both
types of slurries. After storage, it diminishes more for cattle slurries
than for pig slurries; after aeration and anaerobic digestion, it increases, more after aeration and for pig slurries.

Table III. Composition of the slurries before the preparation (C av,g/l),
relative balance of preparation (B %) and comparison of the
slurries before and after preparation (t-test).

Slurries Preparation		pH	MO	%MO/MS	N_T	$N-NH_4$	N_0	$\%N_0/NT$
Cattle	C av	7.00	50.4	82.4	1.94	0.92	1.02	53.3
Storage	B %	-7.6	-2.6	-0.8	-2.1	+8.7	-11.8	-9.4
	t-test	***	n.s.	-	n.s.	***	**	-
Cattle	C av	7.02	49.1	82.4	1.91	0.92	0.99	52.7
Aeration	B %	+9.1	-21.0	-4.2	-9.9	-25.0	+ 4.0	17.1
	t-test	***	***	-	**	***	n.s.	-
Cattle	C av	7.05	45.1	81.1	1.93	1.00	0.93	49.2
Anaer.digest.	B %	+2.7	-31.3	-6.9	0	+8.0	-8.6	-4.2
	t-test	*	***	-	n.s.	***	***	-
Pig	C av	6.99	50.6	77.3	5.23	3.59	1.64	30.5
Storage	B %	-3.1	-4.5	-0.4	-3.4	-0.3	-10.4	-8.2
	t-test	***	**	-	n.s.	n.s.	*	-
Pig	C av	7.00	50.4	77.2	5.21	3.58	1.63	30.6
Aeration	B %	+15.3	-18.7	-2.8	-4.8	-1.7	-11.7	-6.2
	t-test	***	***	-	**	n.s.	*	-
Pig	C av	6.98	50.0	77.3	5.04	3.48	1.56	31.0
Anaer.digest.	B %	+10.3	-23.0	-5.8	0	+14.4	-32.1	-32.3
	t-test	***	***	-	n.s.	***	***	-

t-test: n.s. = not significant
     *   = significant for p 0.05
     ** = significant for p 0.01
     *** = significant for p 0.001
     -    = not calculated

The decrease of the organic matter is the smallest after storage
(< 5 %) and the most important after the anaerobic digestion. However,
we point out a relative lower decrease after the anaerobic digestion
of the pig slurries, since values can be found that reach 40 % in the
practice (5). The decrease of 20 % determined by the aeration is high,
compared to the 10 - 15 % observed under practice conditions by Thalmann (4); the reason for this difference is the intentional strong
intensity of aeration in our conditions.

After storage and anaerobic digestion, the losses of total nitrogen
can be considered as negligible. However, the mineralization of the organic nitrogen is significant, if we consider the decrease of this nitrogen form compared to the increase of the ammoniacal nitrogen. The
significant decreases of the total nitrogen after aeration do not exceed
the 10 % (Thalmann observes a reduction of about 15 % (4)), and are
twice higher for the cattle slurries than for the pig slurries. Whereas
from the point of view of the sum of reactions, the only volatilization
of the ammoniacal nitrogen determines the losses of nitrogen of the aerated cattle slurries, the losses of ammoniac by volatilization in the
aerated pig slurries are continuously compensated by the mineralization
of the organic nitrogen.

Lastly, the comparison of the methods of preparation taken in pairs
(Table IV) shows a characteristic influence of every of them on the different parameters, aside some exceptions; the most important concerns
the total nitrogen content. Their discussion follows from what has already been said.

Table IV Comparison of the three methods of preparation, based on the
composition of the prepared slurries (meaning of t-test: see
Table III)

Types of slurries Compared methods	t-Test				
	pH	MO	N_T	$N-NH_4$	N_O
Cattle					
Storage-aeration	***	***	*	***	**
Storage-anaer. digest.	***	***	n.s.	n.s.	n.s.
Aeration-anaer. digest.	***	***	***	***	***
Pig					
Storage-aeration	***	***	n.s.	n.s.	n.s.
Storage-anaer. digest.	***	***	n.s.	***	***
Aeration-anaer. digest.	***	n.s.	n.s.	***	***

3.2 Efficiency of nitrogen of the prepared slurries

The figure 1 especialy shows for 1980 and for the soil of neutral
reaction, the effects type, method of preparation and amount of slurries
on the four following parameters: dry matter yields, total nitrogen and
nitrate nitrogen contents (for the last only first harvest), and total
nitrogen uptakes of the Italian Rye-grass. The efficiency of the slurries'

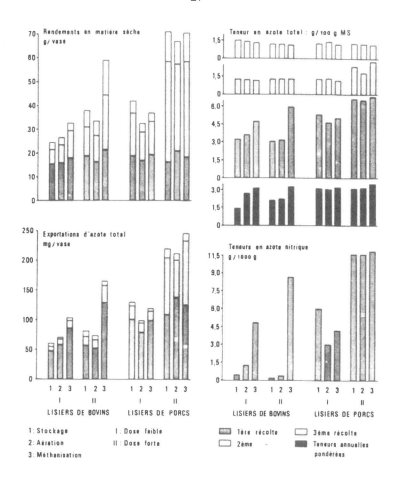

Figure 1 Effet de la préparation des lisiers sur les quatre paramètres observés sur le Ray-grass italien (<u>Lolium italicum</u>) pour l'année 1980 et le sol de réaction neutre.

nitrogen is measured in function of the results obtained by application of mineral nitrogen (nitrate saltpetre 26 %) on the Italian Rye-grass (Table V): the results obtained by the application of slurries are expressed in % of those which were observed by application of mineral nitrogen, this for each of both amounts of nitrogen. Three parameters are considered: yearly dry matter yields, yearly balanced total nitrogen contents (= yearly total nitrogen uptakes: yearly yields) and yearly total nitrogen uptakes.

Three statements should be proposed for these results:
- The efficiency of the pig slurries is higher than that of the cattle slurries, and the difference is more pronounced for the great amount. The lowest efficiency was found for the nitrogen uptake after application of the amount 2 of cattle slurries, especially for the aerated slurries.
- The results verify only the first part of the hypothesis, that is to say the lower nitrogen effect of aerated slurries at the first harvest. However, in the case of the second and the third harvests, the results of the treatments with aerated slurries are equal or smaller than those of the other treatments. For the treatments with aerated slurries, the lowest results of the first harvest are thus not compensated by higher results at the following harvests. The annual values are hence in the main smaller for aerated slurries with respect to the others. It is however to remark that the contribution of the second and third harvests does generally not reach the half of the annual values. It is then hardly possible that the differences observed at the first harvest are compensated - even less passed beyond - by the results of the following harvests.
- Lastly, the relationship between the fraction of ammoniacal nitrogen (mean for the cattle slurries: 46 % and for the pig slurries: 73 %) on the one hand and on the other hand, the yields or the nitrogen uptakes of the Rye-grass, is not obvious, particularly for the pig slurries (Figure 2): even though the effect of the preparation is more pronounced for the cattle slurries than for the pig slurries, the effect of the amounts predominates over the effect of the preparation for the pig slurries.

For yields (field trials),Walther (13) finds nitrogen efficiency values for stocked cattle slurries in the main higher than those presented here; they fluctuate according to the crops: on potatoes, 46-54 %; on winter wheat 69-78 %; on grain maize: 69-110 %. Siegenthaler (11) measures rates of efficiency for stocked pig slurries which vary between 58 and 70 % (mean 65 %) in a pot trial with Italian Rye-grass.

In field trials, the application of aerated cattle slurries determines no increase of the yields with respect to the application of stocked slurries (12, 9). On the contrary, Abele (1) finds higher yields in field and pot trials, mostly significant, after application of aerated cattle slurries, compared to the treatments with stocked slurries. In pot trials, Amberger (2) measures a slightly higher efficiency with aerated cattle slurries. These inconsistent results should be explained by the fact (Abele, personal communication, 1984) that in a light soil(1, 2) the noxious substances of stocked slurries could be prejudicial to the good growth of the roots, because these substances are less adsorbed by the clay-humus complex than in a heavy, well puffered soil (12, 9, 7). The hypothesis of the noxious substances contained in the stocked slurries

Table V. Efficiency of nitrogen of the prepared slurries, calculated in % of the results obtained by application of mineral nitrogen, on three parameters of the Italian Rye-grass (Lolium italicum) (average of three years).

Parameters	Type of slurries	Reaction of the soil	Amount 1			Amount 2		
			Storage	Aeration	Anaer.digest.	Storage	Aeration	Anaer.digest.
Dry matter yields	cattle	neutral	54.4	52.7	66.3	54.5	44.3	65.1
		acid	63.5	60.1	71.0	59.0	49.7	67.5
	pig	neutral	77.1	68.8	73.7	88.6	77.6	86.7
		acid	84.0	77.9	75.8	80.7	76.4	80.2
Total nitrogen contents	cattle	neutral	70.1	68.8	76.5	49.5	48.0	63.3
		acid	75.4	73.8	79.8	49.2	48.9	61.7
	pig	neutral	86.1	86.1	81.4	79.4	73.2	79.8
		acid	91.2	87.1	90.6	75.8	73.9	76.1
Total nitrogen uptakes	cattle	neutral	38.5	37.8	57.0	31.9	24.7	51.3
		acid	50.6	49.0	64.3	37.3	31.4	52.5
	pig	neutral	72.3	66.2	68.3	71.5	60.9	72.3
		acid	81.6	73.8	74.7	68.7	62.4	68.0

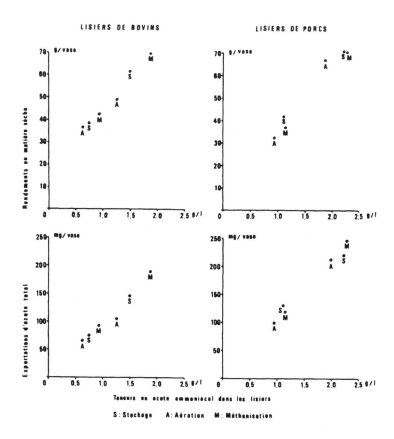

<u>Figure 2</u> Répartition des points sur une droite hypothétique de régression
en fonction des quantités d'azote ammoniacal apportées par les
lisiers de bovins et de porcs après préparation par stockage,
aération et méthanisation, d'une part et, d'autre part, soit les
rendements annuels de matière sèche soit les exportations an-
nuelles d'azote total du Ray-grass italien (<u>Lolium italicum</u>).
Exemple pour les lisiers de bovins: 1981, sol neutre et, pour
les lisiers de porcs: 1980, sol neutre.

has been previously pointed out (10) and is running its course of verification (Ph. D. Thesis, Landesanstalt für Bodenkultur und Pflanzenbau, TU München Freising Weihenstephan).

However, what happened to the organic nitrogen of the used slurries, more precisely of the aerated slurries? Apparently, the mineralization of this nitrogen is considerably slower than expected a priori. In our experiment, as previously pointed out, the Italian Rye-grass has practically stopped to grow after the third cut (average 16 weeks of vegetation), despite a reserve supplying of phosphorus and potassium in the soils. After 40 weeks of incubation, Amberger (2) measures rates of mineralization of the organic nitrogen of 27 and 17 % respectively for cattle and pig slurries.

In short time pot trials (2) with application of slurries in which the ammoniacal nitrogen was almost completely extracted by previous lyophilization, the organic nitrogen of them has only determined an increase of 5 to mostly 10 %, with respect to the control treatment, on the yields and the nitrogen uptake of maize (after 5 weeks of vegetation), of oat (10 weeks) and of Rye-grass (three harvests). On the basis of the present experimentation, it can be deduced that the organic nitrogen of the aerated slurries seems to undergo the same fate as that of the stocked and anaerobic digested slurries: "enter the great pool of soil organic nitrogen, from which the mineralization rate rises usually 1 - 3 % yearly" (2) and is also only hardly available by the plant.

Résumé

Ce travail traite a) de la composition des lisiers de bovins et de porcs après stockage, aération et méthanisation, et b) de l'efficacité de l'azote des lisiers préparés.

a) Les lisiers aérés contiennent le moins d'azote minéral (donc le plus d'azote organique), les lisiers stockés occupent une position intermédiaire et les lisiers méthanisés ont les teneurs les plus élevées en azote minéral.

b) L'efficacité de l'azote est mesurée dans des essais en vases de végétation avec du Ray-grass italien (Lolium italicum) récolté trois fois par an. Celle des lisiers aérés est la plus faible, non seulement pour la première récolte (effet direct) mais aussi pour les suivantes (arrière-effet décevant). On en déduit que l'azote organique des lisiers aérés subit le même sort que celui des autres: "entrer dans le grand réservoir de l'azote organique des sols, dont le taux de minéralisation s'élève habituellement à 1 - 3 % par année" (2), et n'est, lui aussi, que peu disponible pour la plante. Enfin, en fonction des méthodes de préparation, la relation entre la part d'azote ammoniacal dans les lisiers et les paramètres observés sur le Ray-grass, n'est pas évidente. Au niveau de la réponse de la plante, les effets type et dose de lisiers prédominent sur l'effet préparation, en particulier pour les lisiers de porcs.

REFERENCES

(1) ABELE,U.: Untersuchung des Rotteverlaufs von Gülle bei verschiede-
      ner Behandlung und deren Wirkung auf Boden, Pflanzenertrag und
      Pflanzenqualität. Abschlussbericht, Inst. biol.-dyn. Forsch.,
      Darmstadt, 205 S. (1976).

(2) AMBERGER, A., VILSMEIER, A. und GUTSER, R.: Stickstofffraktionen
      verschiedener Güllen und deren Wirkung im Pflanzenversuch. Z.
      Pflanzenernähr. Bodenk., 145, 325-336 (1982).

(3) BESSON, J.-M.: Behandlung von Jauche und Gülle. In: Oekologische
      Landwirtschaft, Kongressbericht "Grünes Forum Alpbach 1980". Wag-
      ner'sche Universitätsbuchhandlung, Innsbruck, 87-120 (1981).

(4) BESSON, J.-M.: Güllebehandlung, Lagerung, Belüftung und Methan-
      gärung: Wirkungen auf einige Gülleinhaltsstoffe. In: "Fragen der
      Güllerei", 7. Arbeitstagung Gumpenstein, 29.9.-2.10.1981. Verlag
      und Druck der Bundesversuchsanstalt für alpenländische Landwirt-
      schaft, Gumpenstein, I. Band, 29-60 (1981).

(5) BESSON, J.-M., LEHMANN, V., ROULET, M. et WELLINGER, A.: Influence
      de la méthanisation sur la composition des lisiers. Rev. Suisse
      Agric., 14, 143-151 (1982).

(6) BESSON, J.-M., LEHMANN, V., ROULET, M. et EDELMANN, W.: Comparaison
      de trois traitements des lisiers en conditions expérimentales con-
      trôlées: stockage, aération, méthanisation. Rev. Suisse Agric., 14,
      327-335 (1982).

(7) BESSON, J.-M., DANIEL, R. et LISCHER, P.: Comparaison de l'effet de
      l'azote dans des lisiers de bovins et de porcs sur le Ray-grass
      italien (Lolium italicum), après préparation selon trois méthodes:
      stockage, aération et méthanisation. Rech. Agron. Suisse, 23, 249-
      268 (1984).

(8) BIOGASHANDBUCH: Autorenkollektiv. Verlag Wirz AG, Aarau, ca 250 S.,
      im Druck (1984).

(9) BUCHGRABER, K.: Vergleich der Wirksamkeit konventioneller und alter-
      nativer Düngungssysteme auf dem Grünland, hinsichtlich Ertrag, Fut-
      terqualität und Güte des Pflanzenbestandes. Diss. Inst. Pflanzen-
      bau-Pflanzenzücht., Univ. Bodenkultur, Wien, 213 S. (1983).

(10) NEBIKER, H.: Neues Verfahren zur Aufbereitung von Flüssigdünger.
      Schweiz. Landw. Monatshefte, 52, 523-531 (1974).

(11) SIEGENTHALER, A. und BOLLIGER, R.: Untersuchungen über die Wirksam-
      keit des Stickstoffes von Schweinegülle und Geflügelmist in einem
      Gefässversuch. Schweiz. Landw. Monatshefte, 58, 523-531 (1980).

(12) THALMANN, H. Güllebehandlung, Lagerung, Belüftung und Methangärung:
      Wirkungen belüfteter und unbelüfteter Gülle im Feldversuch. In:
      "Fragen der Güllerei", 7. Arbeitstagung Gumpenstein, 29.9. - 2.10.
      1981. Verlag und Druck der Bundesversuchsanstalt für alpenländische
      Landwirtschaft, Gumpenstein, I. Band, 61-78 (1981).

(13) WALTHER, U.: Pflanzen- und umweltgerechter Einsatz von Gülle im Ak-
      kerbau. In: "Flüssigdüngung", 12. SVLT-Vortragstagung. 9.12.1983,
      Schönbühl-Urtenen, 6. S. (1983).

LONG-TERM EFFECTS OF THE LANDSPREADING OF PIG AND CATTLE SLURRIES ON THE
ACCUMULATION AND AVAILABILITY OF SOIL NUTRIENTS

P. SPALLACCI[1] and V. BOSCHI[2]
Ministero dell'Agricoltura e delle Foreste (Italy)
[1]Istituto sperimentale per lo studio e la difesa del suolo, Sezione di
Chimica del suolo, Firenze
[2]Istituto sperimentale agronomico, Sezione di Modena

Summary

Long-term experiments were conducted in plots and in lysimeters with
pig and cattle slurry applications, repeated for several years
(from 3 to 5) on different soil types. The soil contents of total N,
organic C, available P and exchangeable K were measured at various
times: a) beginning of the trials; b) end of the slurry treatment
period; c) end of period of residual effects. High enrichments of
nutrients were found at the end of the slurry application period and
optimum crop yields were obtained during this time. In the subsequent
period of residual effects, in order to obtain high yields, additional
N fertilizations were necessary, whereas the available P and exchangeable K contents were sufficient for crop requirements for 2-3 years.
After such period, the P for pig slurry and K for cattle slurry
indicate still high levels of enrichment.

1. INTRODUCTION

The agronomic utilization of animal slurry still remains the most
natural and economic solution, notwithstanding attempts at disposal by
means of biochemical depuration (activated sludge), or of energy production
(biogas). The choice for the spreading on the soil made by the EEC Commission at the beginning of the Seventies results today as fully confirmed,
and already since some time has been accepted by the Italian legislation
in respect to the protection of water resources.

The Ministry of Agriculture promoted the first studies in Italy on
this subject, which were then coordinated in the EEC Programme "Effluents
from livestock" of 1976-1979. The results of our research, conducted within
the ambit of the Community programme (Contract n.235), are published in the
reports of the various Seminars and Workshops organized mostly by EEC until
1982 (1,2,3,4,5,6,7).

The long period of the study carried out enables us today to present
some results of the trials set up at the Sezione di Modena of the Istituto
Sperimentale Agronomico, to assess the long-term effects of the spreading
of slurry on the soil, and specifically:

1) treatments for 5 consecutive years with pig slurry on clay soil tested from 1970 to 1983 (experiment 1);
2) treatments for 4 consecutive years with pig slurry on three types of soil tested from 1976 to 1983 (experiment 2);
3) treatments for 3 consecutive years with cattle slurry on three types of soil tested from 1978 to 1983 (experiment 3).

2. EXPERIMENTAL RESULTS

2.1 Experiment 1

The field tests were run on a clay soil similar to that used for the lysimeter experiment, shown hereafter (Table I). Following an experimental design with increasing rates of pig slurry applied to large plots, the treatments were repeated for 5 consecutive years (1970–74). The two highest rates (750 and 1500 m^3 ha^{-1} $year^{-1}$) were included to test an hypothesis of slurry disposal. The effects on the yield and quality of crops and on some characteristics of the soil, in the 5–year period 1970–74 and subsequently up to 1979, have already been presented in other papers (1,2,5). Starting in 1975 the slurry treatments were suspended with the aim of studying residual effects on crop productivity and the fertility evolution of the soil. From 1975 to 1983, the following crops were grown: maize, sorghum and sugar beet in the first three years; lucerne from 1978 to 1982; barley in 1983.

The contents of the main soil nutrients (0–40 cm layer) in 1970, in 1974 and in 1983 are given in Figure 1. After the 5–year period 1970–74 of repeated treatments with slurry, the soil resulted considerably enriched: in particular, total P, available P and exchangeable K show the greatest differences.

After nine years without any manurial treatment, the nutrient contents of the soils which were previously enriched decreased more extensively where the enrichments had been greater, with the exception of total P which remained unchanged. Furthermore, a parallelism is found between total N and organic C, whereas a levelling of the exchangeable Na to the values of the control is noted. The strong reduction of available P and exchangeable K are due both to crop absorption and also to the effects of microbiological and physical–chemical processes. Finally it is to be reported that nine years after the slurry treatments ceased, in the soil it can be observed amounts of nutrients considerably greater than normal only for the highest rates of slurry which, as has already been said, are not suitable for fertilizing practice.

2.2 Experiment 2

The tests were made in lysimeters (1x1x1 m) on three different soils treated with increasing rates of pig slurry, with the aim of identifying the best amounts, in respect to the yield and quality of crops and to the resulting effects on the soil and on the percolation waters. From 1976 to 1979, annual applications of pig slurry were made to the following

TABLE I. Physical and chemical composition of the different soils used in lysimeter experiments on pig slurry (1)

ANALYSES		Sandy loam (Xerochrept)	Sandy clay (Fluventic Xerochrept)	Clay (Vertic Xerochrept)
Physical analyses (2)				
Sand (2-0.02 mm)	%	56.5	63.7	10.9
Silt (0.02-0.002 mm)	%	32.6	20.5	47.5
Clay (<0.002 mm)	%	10.9	15.8	41.6
Bulk density	g/cm^3	1.32	1.26	1.25
Total porosity	%	47.2	49.2	52.4
Water infiltration rate	cm/h	20.7	16.1	19.0
Water retentivity { pF 2.54	%	21.8	21.5	34.1
Water retentivity { pF 4.19	%	12.2	13.1	22.1
Chemical analyses (3)				
pH (in H$_2$O)		7.2	7.9	8.0
pH (in KCl)		6.3	7.3	7.4
Conductivity at 20° (1:2.5)	mmho/cm	0.149	0.175	0.203
Total CaCO$_3$ (Scheibler)	%	0.6	14.4	14.6
Active Ca CO$_3$ (Drouineau)	%	0.6	3.2	10.0
Total N (Kjeldahl)	‰	1.25	1.11	1.44
Organic C (K$_2$Cr$_2$O$_7$)	‰	9.30	7.03	9.77
C/N ratio		7.4	6.4	6.8
Total P	‰	0.550	0.515	0.585
Extractable P (0.5 N NaHCO$_3$, pH 8.5)	ppm	15.7	5.2	7.0
Exchangeable K (1N NH$_4$OAc)	ppm	105	147	290
Soluble K (1:2.5)	ppm	10	25	14
Exchangeable Na (1N NH$_4$OAc)	ppm	15	0	25
Soluble Na (1:2.5)	ppm	13	16	22
CEC (BaOAc, pH 7)	meq/100 g	20.1	20.2	25.7

(1) samples made before the start of trials, at 0-25 cm depth;
(2) data expressed on oven-dry basis; (3) data expressed on air-dry basis.

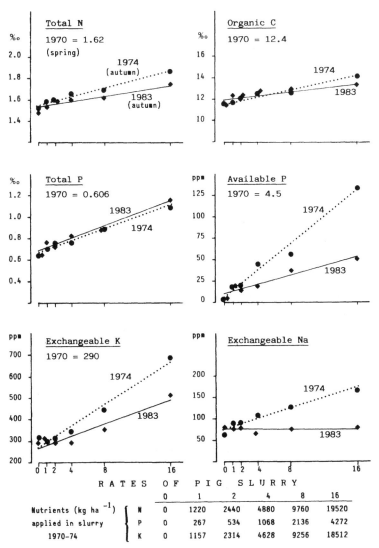

Fig. 1. Contents of main nutrients in a clay soil treated with increasing rates of pig slurry: comparison of the beginning of the trials
(1970) with the end of 5-year period of annual slurry applications
(1974) and the end of further 9-year period without manurial treatments
(1983).

crops: 1976, forage sorghum; 1977, grain maize; 1978, wheat+forage maize;
1979, grain sorghum. In the subsequent 3-year period (1980-82), cockfoot
was grown without any manurial treatment. Results on the subject have
already been published (2,3,4,6), and further details on materials and
methods can be obtained from these papers. Here it is desired to assess,
on the one hand, the entity of the nutrient accumulation following
treatments repeated over four years and, on the other hand, the evolution
of the residual fertility in the subsequent three years.

The characteristics of the three soils studied for 0-25 cm depth are
reported in Table I, while the nutrient amounts applied to the soil and
the average composition of the slurry are given in Table II. The main
effects of the slurry application on the chemical-nutritional characteristics of the soils are represented in Figure 2. After four years of
slurry applications, all soils were enriched of total N and organic C, the
greatest increase being evident in the sandy clay soil. As for the accumulation of N and C, a substantial parallelism was confirmed; however for
higher rates of application a tendency to the increase of the C/N ratio
was noted. The enrichment in the three soils were very high either for
total P (respectively 0.38 - 0.40 - 0.34 ‰ P are the differences between
maximum rate and control) or for available P (respectively 8 - 17 - 16 are
the ratios between maximum rate and control). Moreover, the maximum
content of available P in each soil resulted slightly lower when higher
contents of clay and/or limestone were present. The increases in exchangeable K were considerable in all three soils; but if compared with the
available P, they appear much lower (respectively 2.1 - 1.8 - 1.3 are the
ratios between maximum rate and control), as is evident also from the
trends of the regression lines (Fig. 2).

The same analyses, repeated after a 3-year period of cockfoot meadow
without any fertilization, have shown a general decrease in nutrient from
the levels previously attained. It can also be confirmed that the deduction
was sensitive when the enrichments were great, both in function of the
rates of slurry and in reference to the type of soil. The only exception
to this general trend was that of the clay soil in which the exchangeable K
and the organic C did not show decreases, so that an increase in the
C/N ratio was noted. Despite the decreases encountered after the 3-year
period without fertilizers, the enrichment of total N, organic C and
available P remained, however, high in all soil types, although at a
slightly lower level for N and C in the sandy loam soil. The exchangeable K
was reduced practically to the initial levels in the two soils with lesser
clay contents; whereas it still presents some enrichment in the more clay
soil, as related to the increasing rates of slurry.

2.3 Experiment 3

The tests were run on small plots (2x2 m) limited laterally by con-
crete plates and filled to a thickness of 50 cm, with three types of soil
similar to those in Table I, except for some contents of nutrients
owing to previous crop rotations and fertilizer additions. For the first

Table II. Organic matter and nutrients supplied by pig slurry applications to various crops in lysimeter trials during 1976-79 and average composition of the slurry.

Constituents	Amounts applied in rate 1 of slurry kg ha^{-1}						Average composition of slurry % wet-weight
	Forage sorghum 1976	Grain maize 1977	Wheat 1977-78	Forage maize 1978	Grain sorghum 1979	Annual average	
Dry matter	3074	3033	2827	1975	4366	3819	3.39
Organic matter	1978	2107	2199	1464	2861	2652	2.36
Total N	440	280	218	176	308	355	0.316
NH$_3$-N	266	180	126	112	183	217	0.193
Total P	126	89	70	68	110	116	0.103
Total K	381	311	140	92	146	267	0.238
Na (from chloride)	146	96	42	30	36	87	0.078
Total Ca	85	75	72	69	121	105	0.094
Total Mg	48	45	37	32	32	58	0.049

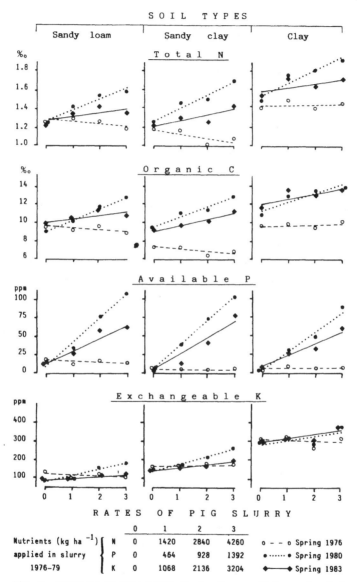

Fig. 2. Contents of main nutrients in three soil types treated
with increasing rates of pig slurry: comparison of the beginning
of trials (1976) with the end of 4-year period of annual slurry
applications (1980) and the end of further 3-year period without
manurial treatments (1983).

three consecutive years (1978–80), increasing doses of cattle slurry were
applied, whereas during the subsequent two years, the residual effects were
utilized. Silage maize was cropped throughout the all period. The amounts
of applied nutrients and the average composition of the slurry are given
in Table III.

The contents of the main soil nutrients (0–25 cm layer) at the
beginning and at the end of the 5–year period (1978–83) are presented in
Figure 3. The comparison of the data of 1978 with the ones of 1983 show
that the initial level of N and C in the sandy loam soil was kept constant
only by applying the maximum rate of slurry. In the clay soil, instead,
an amount equal to a third of the maximum rate was sufficient. This result
confirms the grater mineralizing capacity of the light soils, in this
case particularly marked since we are dealing with a soil rich in ferric
oxides. On the other hand, five years with continous maize cropping could
have accentuated the oxidation processes in the soil. In 1983 the available
P contents resulted considerably increased by the increasing amounts of
slurry, more in the sandy loam soil – despite its greater natural reserve –
than in the other two soils. A slight decrease in the P availability was
evidenced in the control, and the lowest amount of slurry was sufficient
to maintain the levels present in the soils at the beginning of the trials.
The exchangeable K strongly increased as a consequence of the increasing
amount of slurry: the enrichment from 1978 to 1983 was almost identical
in the three soils, despite their considerable different natural content,
related to their mineralogical characteristics. In the absence of manurial
additions the K contents were slightly lowered; but a quantity of slurry
equal to half the lowest rate applied would already have been sufficient
to maintain unchanged the exchangeable K levels in the three soils.

3. CONCLUSIONS

The applications of pig slurry in increasing rates, and repeated for
several years, determined an enrichment of total N and organic C in the
soils. If compared with the rates applied, the increases found appear very
low, but nevertheless they hold a high practical interest for the achieve-
ment or maintenance of a good level of organic matter in the soil. The
available P and exchangeable K manifested much more marked increases than
the N and the C. In particular, the P had a greater increase in the soils
containing less clay and/or limestone.

In the three cropping years for the utilization of residual effects,
the level of all   nutrients was reduced, even more intensively where
the accumulations were greater. Nevertheless, for total N, organic C and
available P, substantial enrichment persisted; whereas for exchangeable K,
a certain levelling to the initial contents was the main feature. These
results find an explanation in the K amounts supplied by slurry, relatively
low with respect to those of N and mainly of P. Supplying too high amounts
of nutrients, as in the disposal practice, the enrichment of a clay soil
was high for all nutrients, also after a 9–year period without manurial
treatments.

Table III. Organic matter and nutrients supplied by cattle slurry applications to silage maize in 1978–80 and average composition
of the slurry.

Constituents	Amounts applied in rate 1 of slurry				Average composition of slurry
	1978	1979	1980	Annual average	
	kg ha^{-1}				% wet weight
Dry matter	3717	5278	5108	4701	6.27
Organic matter	2855	4033	4036	3641	4.85
Total N	232	270	250	250	0.334
NH_3-N	123	148	139	137	0.182
Total P	38	61	52	50	0.067
Total K	214	427	351	330	0.440
Na (from chloride)	50	114	96	87	0.116
Total Ca	79	105	70	85	0.113
Total Mg	33	53	37	41	0.055

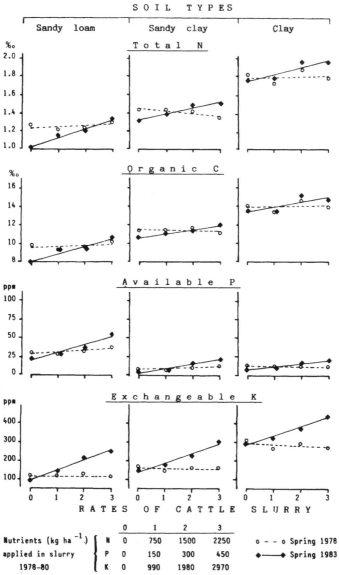

Fig. 3. Contents of main nutrients in three soil types treated
with increasing rates of cattle slurry: comparison of the beginning
of trials (1978) with the end of 5-year period (1983) of which 3-year
period of annual slurry applications and 2-year period without
manurial treatments.

As far as cattle slurry is concerned, the applications repeated for three years (this period was then followed by two years of residual effects) caused particularly high levels of exchangeable K; but nevertheless, the increases of total N, organic C and available P were also remarkable. This agrees with the applied amounts of N and P, relatively low with respect to the quantity of K, of which cattle slurry is particularly rich.

In the study of the N balance in lysimeters with pig slurry (6,7), recommended amounts of slurry, supplying a total N input of 220-300-500 kg ha^{-1} year^{-1} respectively for the three soil types considered, were established. These amounts, defined for pig slurry (but with an increase of 10-20% valid also for cattle slurry) and repeated for 3-4 consecutive years, resulted sufficient for the nitrogen requirement for high crop yields. In the years of utilization of the residual effects, instead, high yields required addition of N fertilizer (2,7): greater addition was necessary for pig slurry than for cattle slurry, obviously taking into consideration the crop and the type of soil.

Also with the application of amounts thus calculated, the surplus of available P and exchangeable K in the soil at the end of the treatment period with slurry resulted high. These reserves were sufficient for the crop requirements for at least 2-3 subsequent years and for maintaining the P and K availability at greater levels than the initial ones in different soils. After such period, the soil contents of available P (if pig slurry is used) and exchangeable K (if cattle slurry is used) still remained at enrichment levels, which must be taken into account for the correct dosage of the subsequent fertilizer applications.

4. REFERENCES

1. BOSCHI V., SPALLACCI P., MONTORSI M.: The agronomic utilization of pig slurry: effect on forage crops and on soil fertility. Summary of a five-year investigation. In: Voorburg J.H. (Ed.) "Utilization of manure by land spreading". CEC, Luxembourg, 1977, EUR 5672e, 105-118

2. SPALLACCI P., BOSCHI V.: Spreading of pig and cattle slurries on arable land: lysimeter and field experiments. In: Gasser J.K.R. (Ed.) "Effluents from livestock". Appl. Sci. Publ., Barking, Essex, 1979, EUR 6633 EN, 241-274.

3. SPALLACCI P.: a) Nitrogen uptake by crops manured with pig slurry, 178-180; b) Nitrogen losses by leaching on different soils manured with pig slurry, 284-288; c) Nitrogen accumulation in different soils after

repeated applications of pig slurry, 377–379. In: Brogan J.C. (Ed.)
"Nitrogen losses and surface run–off from landspreading of manures".
Nijhoff/Junk Publ., The Hague (NL), 1981, EUR 6898.

4. SPALLACCI P.: Influence de l'azote de lisier de porc sur le rendement
du blé d'hiver, l'exportation par la plante et la perte par lessivage.
In: "Devenir de l'azote dans la fertilisation N des blés d'hiver.
Influence de l'alimentation hydrique". INRA–CEE, Toulouse, 1981, 101–112.

5. BOSCHI V., SPALLACCI P., MONTORSI M.: Aspetti agronomici dell'utilizzazione dei liquami di allevamenti suini. In: "Inquinamento del terreno.
Somministrazione al terreno degli effluenti da allevamenti intensivi
zootecnici". CNR, Collana PF "Promozione della qualità dell'ambiente"
Pisa, 1982, 9–24.

6. SPALLACCI P.: Bilancio dell'azoto nella somministrazione dei liquami di
suini al terreno. Ibidem, 33–49.

7. SPALLACCI P.: A survey of the most recent research results in Italy on
the fertilization of soil with liquid manure. EEC Workshop under
Programme "Effluents from intensive livestock", Hørsholm (DK), 29 November–1 December 1982 (in press).

RELATIONSHIPS BETWEEN SOIL STRUCTURE AND TIME OF
LANDSPREADING OF PIG SLURRY

M. Pagliai, M. La Marca and G. Lucamante
Institute for Soil Chemistry, C.N.R.
Via Corridoni 78, 56100 Pisa (Italy)

SUMMARY

A field test was established in 1980 on a silty clay soil to study
various aspects of soil porosity related to soil structure following
the landspreading of pig slurry at three different times (February,
June and October) each year. Porosity measurements were carried out
on large thin sections of undisturbed soil samples by means of
electro-optical image analysis (Quantimet 720).
The landspreading of pig slurry increased significantly the total
porosity and modified the size distribution, the shape and the
arrangement of pores. Such modifications were the result of an
increase of soil aggregates following the addition of pig slurry.
This improvement was proportional to the application rates of
slurries, but it was different according to the different times of
landspreading. At the beginning of the experiment the application of
June was the most efficacious while the application of October did
not show differences with respect to the control. In the last year
the application of June still showed the best result but the
application of October also showed a significative improvement with
respect to the control. The effect of the February application was
intermediate between the other two.

1. INTRODUCTION

Soils in the process of continous cultivation are very often
deficient in organic matter, particularly in Italy, where the decomposition rate of soil organic matter is high firstly because of the climate
and secondly because the organic matter returned from crop residues is
less than that returned to the soil from a full vegetation cover.
Moreover, the use of heavy machinery often causes soil damage due to
excessive compaction.
The agronomic utilisation of animal slurries may induce changes in
the physical condition and chemical composition of the soil and knowledge
on these processes is considered necessary in order to obtain dependable

application systems. Figures in the effects of application of sludges on soil are usually concerned with fertility and pollution and rarely with physical properties associated with the structure of the soil, an important parameter of soil fertility and crop yield. Porosity, pore size distribution and pore shape give important data on soil structure, because pores determine various physical properties important to plants.

This study was undertaken to determine the effects of pig slurry, landspreaded at different times, on soil structure and particularly on soil porosity under natural field conditions. Measurements were carried out analysing soil thin sections by an electro-optical image analysis. The feasibility of using such an approach to measure and characterize pores in soil has been demonstrated by many authors (1,5,7,8). Micromorphological investigations were also carried out in order to study the influence of pig slurry on surface soil crust formation.

2. THE EXPERIMENTS

A long-term field study was established in October 1979 on a silty clay soil of the experimental farm S. Prospero of the "Istituto Sperimentale Agronomico", Section of Modena. The livestock effluents used were pig slurry and the trials were carried out on large plots planted to corn. The soil (0-10 cm) contained 42% clay, 45% silt, 13% sand and 2% organic matter. The pH was 8.1 in water.

Pig slurries were surface applied and the following treatments were compared:
1) Control (C)
2) 100 m^3/ha of pig slurry applied the 1st February   (T1-100)
3) 200 " " " " " " " "   (T1-200)
4) 300 " " " " " " " "   (T1-300)
5) 100 " " " " " " " June   (T2-100)
6) 200 " " " " " " " "   (T2-200)
7) 300 " " " " " " " "   (T2-300)
8) 100 " " " " " " " October   (T3-100)
9) 200 " " " " " " " "   (T3-200)
10) 300 " " " " " " " "   (T3-300)
11) 400 q/ha of farmyard manure applied before ploughing (FYM)
These treatments were repeated each year.

3. METHODS

Five undisturbed and oriented soil samples were collected from the Ap horizon of each plot at the end of June of each year. Samples were carefully packed to avoid any breakdown, air dried at room temperature, impregnated with a polyester resin and made into large (6x6 cm) thin section of 20 um of thickness (4). A photographic procedure was used in order to separate pores from mineral grains (3). Each photograph was then analyzed by an image analyzing apparatus (Quantimet 720).

In this instrument the image formed by a thin section under the microscope or by a photograph under the epidiascope is scanned by a plumbicon television camera and displayed on the screen of a monitor. The video signal is passed to a detector where 500,000 points on the image are individually analyzed for their gray level. Pores are measured by setting the instrument to detect the corresponding gray level. In these investigations measurements include the number and area of pores.

Surface soil crusts were examined on thin sections, prepared from undisturbed and oriented samples taken from the surface of the soil, by a Leitz Orthoplan microscope at 40 magnifications.

4. RESULTS AND DISCUSSION

In this paper there is the discussion of results obtained in the first four years (1980-1983). Mean values of total porosity, expressed both as a percentage of the total area and as the number of pores per thin section are reported in tables I and II, respectively.

For a better understanding of data reported in this table it is necessary to underline that according to this micromorphometric method, a soil is considered very dense with a total porosity less than 5%, dense with a total porosity of 5-10%, moderately porous 10-25%, highly porous 25-40% and extremely porous, with a total porosity greater than 40%.

Table I - Effects of treatments on total soil porosity expressed as a percentage of the total area of thin section occupied by pores. Data are the mean of five repetitions.

Treatment	Total porosity (%)*			
	1980	1981	1982	1983
Control	22.6 a	19.1 a	18.5 a	17.1 a
T1 - 100	22.5 a	21.3 a	23.7 b	28.5 b
T1 - 200	24.7 b	22.4 ab	25.2 b	31.0 b
T1 - 300	31.0 c	25.3 b	29.4 c	38.2 c
T2 - 100	33.4 c	29.6 c	35.1 d	39.5 c
T2 - 200	42.0 d	38.3 d	40.0 c	44.3 d
T2 - 300	45.1 e	39.2 d	46.2 f	47.2 c
T3 - 100	21.4 a	18.4 a	22.1 ab	26.4 b
T3 - 200	22.3 a	19.5 a	25.4 b	28.3 b
T3 - 300	20.0 a	20.3 a	26.3 b	30.1 b
FYM	25.4 b	22.5 ab	28.1 c	29.8 b

* Means in a column followed by the same letter are not significantly different at 0.05 level employing Duncan's Multiple Range Test.

Table II - Effect of treatment on soil porosity expressed as total number of pores per thin section. Data are the mean of five repetitions.

Treatment	Total number of pores*			
	1980	1981	1982	1983
Control	716 a	700 a	655 a	610 a
T1 - 100	790 a	803 a	859 b	1090 b
T1 - 200	820 ab	848 ab	930 b	1303 b
T1 - 300	853 ab	921 b	1101 c	1412 b
T2 - 100	969 b	1034 c	1213 d	1704 c
T2 - 200	1284 c	1528 d	1905 c	2019 d
T2 - 300	1816 d	1973 c	2102 f	2155 c
T3 - 100	723 a	779 a	838 b	1088 b
T3 - 200	648 a	805 a	901 b	1215 b
T3 - 300	751 a	804 a	920 b	1400 b
FYM	841 ab	1027 c	1312 d	1703 c

* Means in a column followed by the same letter are not significantly different at 0.05 level employing Duncan's Multiple Range Test.

Soil samples taken in the plots in which the addition of pig slurry was made the 1st June (about one month before the sampling) showed always the highest values of total porosity and the increase with respect to the control was proportional to the addition rate of pig slurry. A high proportion of total porosity was formed by biopores originated both by soil fauna and by carbon dioxide liberated by biological activity, confirming that the maximum evolution of carbon dioxide was reached in the first few weeks following the addition of pig slurry to soil (11) expecially in June when the temperature and humidity were optimal for biological activity. Annual variations were not significative. In the first year (1980) a sampling was also made at the end of September and the total porosity showed a decrease with respect to the sampling of June not only in the treated plots but also in the control ones (6). This behaviour was ascribed to the natural compactness of the soil.

A quite different trend during the four years of investigations was found in samples from plots treated in October (T3) and in February (T2). In the first two years (1980, 1981) the samples from plots treated in October did not show any differences with respect to the control plots. On the contrary in the last two years (1982, 1983) the total porosity of plots treated in October showed a significant increase with respect to the control. This increase was porportional to the addition rate of pig slurry and in 1983 values of total porosity increased also with respect to 1982.

The effect of February application was intermediate between those of October and June. In the first two years (1980, 1981) only in samples from plots treated with the lower addition rate of pig slurry the total porosity did not show significative differences with respect to control plots. On the contrary in 1982 and 1983 also the lowest addition rate increased the total porosity significantly with respect to the control.

Comparing the effect on soil porosity of pig slurry and farmyard manure it was found that after four years of treatment all the addition rates of June and the highest rate of February (T1-300) seemed more efficacious than farmyard manure in increasing total porosity. The effect of all others addition rates was similar to that of the farmyard manure.

The number of pores showed the same trend of the percentage area, however after four years all treatments seemed more efficacious in improving the number of pores than their area percentage. In fact, in 1983 the ratio of porosity percentage of T1-300, T2-300, T3-300 to the control was 2.2, 2.8 and 1.8 respectively, while for the pore number it was 2.3, 3.5 and 2.3, respectively. At the beginning of the experiment these ratios were almost the same for porosity percentage and the number of pores. This means that the landspreading of pig slurry and also farmyard manure modified the pore size distribution by reducing the proportion of large pores (larger than 500 μm of equivalent pore diameter - e.p.d.-) and increasing the pores less than 500 μm e.p.d.. From the agronomic point of view, this is a positive result because, according to Greenland (2), for upland crops, adequate storage pores (0.5-50 μm e.p.d.) which retain available water as well as adequate transmission pores (50-500 μm e.p.d.) are necessary. Moreover, roots grow easily only into pores ranging from 100-300 μm.

The variations of soil porosity directly reflected the soil structure. In fact, the microscopic observation revealed differences between the control and treated plots. In samples of control plots the presence of large rather compact soil aggregates could be noticed (Fig.1). These large aggregates were usually surrounded or separated by a few large cracks which represent the highest proportion of the percentage area occupied by all pores. In large areas of the control plots these soil aggregates had a diameter greater than 6-8 cm and a massive interior structure, as reported in Fig.2, could be noticed. The presence of the latter kind of soil aggregate in control plots seemed to increase after the four years of investigations with respect to the beginning.

In samples of plots treated in June the soil structure always showed great differences with respect to the control. The high porosity and above all the high number of pores produced an angular or subangular blocky microstructure (Fig.3) which was better for plant growth (9).

In samples of plots treated in February and in October in the first year the soil structure did not show significative differences with respect to control samples (Fig.1 and 2). On the contrary, after four years the structure of these soil samples was more similar to the samples

50

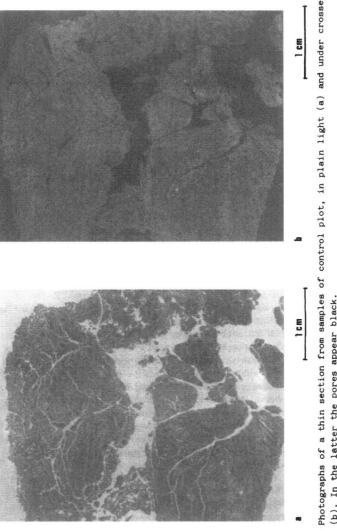

Fig. 1 – Photographs of a thin section from samples of control plot, in plain light (a) and under crossed nicols (b). In the latter the pores appear black.

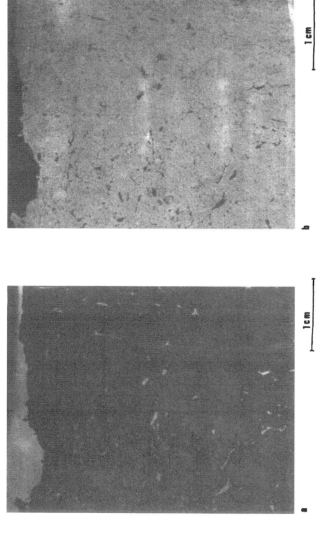

Fig. 2 – Photographs of a thin section from samples of control plot in plain light (a) and under crossed nicols (b). A massive structure is very evident.

52

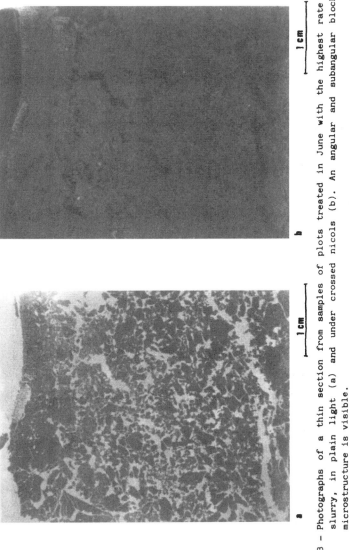

Fig. 3 - Photographs of a thin section from samples of plots treated in June with the highest rate of pig slurry, in plain light (a) and under crossed nicols (b). An angular and subangular blocky soil microstructure is visible.

of plots treated in June than to the control samples. Fig. 4 shows an angular to subangular blocky microstructure of samples taken in 1984 in plots treated in October. This figure shows very few differences with respect to Fig.3 except for the presence of some soil aggregates larger and more compact than those in samples of plots treated in June (Fig.3).

The effect of the application of farmyard manure on soil structure was similar to that of the landspreading of pig slurry in October and February.

The application of pig slurry to agricultural land induces modifications not only on soil porosity but also on surface soil crusts. These specific physical modifications in the top-soil can be caused by natural events like raindrop impact and the following drying process, and they consist in the formation of hard thin layers widespread especially in the soil of arid and semiarid regions. Their thickness usually ranges from 0.5 mm to 5 cm. When dry, these features are more compact, hard and brittle than underlaying soil material and decrease both the size and the number of pores, reducing, in this way, water and air permeability. From the agronomic point of view the most important disadvantages of soil crusts are the influences they have on seedling emergence and water penetration. The latter leads, as a consequence, to an increase of surface runoff.

These surface soil crusts were very common in the control plots, whereas in plots treated with pig slurry they were much less developed. The effect of pig slurry in preventing or reducing soil crusts could be ascribed both to the organic matter which, as it is well known, increase the aggregate stability and to the organic materials, such as straw fragments, which remain in the soil surface reducing the direct action of raindrop impact on surface soil aggregates. These organic materials could also break the surface soil crusts reducing their compactness and improving water intake rates.

The pig slurry landspreadings of February and June were more efficacious in reducing soil crust formation than the landspreading of October even though in the last two years in the plots treated in October the surface crusts were less evident than in the control plots. The farmyard manure also reduced the formation of soil crusts with respect to the control and its action was similar to that of the highest rate of pig slurry landspreaded in October.

5. CONCLUSIONS

For all soil parameters considered in this study a generalized improve of soil structure in samples treated with pig slurry and farmyard manure was observed. The highest improvement of soil conditions was always found in plots where the landspreading of pig slurry was made in June. In plots where the pig slurry landspreading was made in October soil conditions did not show significative differences respect to the control plots at the beginning of the experiment, but after four years the improvement of soil conditions were evident and more similar to those of

54

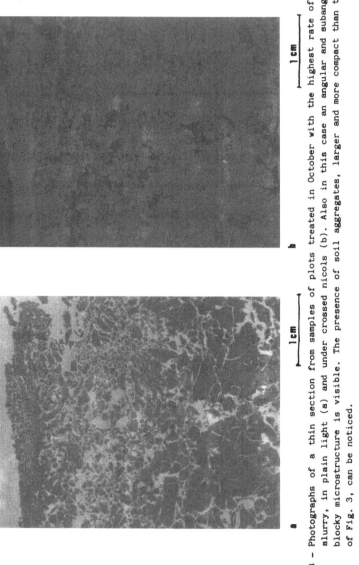

Fig. 4 – Photographs of a thin section from samples of plots treated in October with the highest rate of pig slurry, in plain light (a) and under crossed nicols (b). Also in this case an angular and subangular blocky microstructure is visible. The presence of soil aggregates, larger and more compact than those of Fig. 3, can be noticed.

plots treated in June than to the control plots. The effect of the February landspreading was intermediate between the other two. These data were in agreement with agronomic results: in fact, the crop yield was always higher in plots treated in June. In plots treated in February and October the crop yield increased year by year and in 1983 was quite similar to that of plots treated in June (Spallacci, personal communication).

The increase of both total porosity and pores smaller than 500 μm e.p.d. in treated plots was associated with an improved soil structure and such an increase leads, as a consequence, to a rise in both the level of water available to plants and the pores needed by feeding roots. Besides, it has been found that damage to soil structure can be recognized by decreases in the proportions of pore space present in storage and transmission pores (2). From the microscopic observations it is also possible to assess that organic materials play an important role in preventing soil crust formation.

Data found in this experiment confirm previous results obtained in another experimental fields of the same farm planted to wheat and barley (10). Also in those cases the application to the soil of pig slurry seemed to improve soil porosity and to reduce the formation of surface soil crusts.

6. REFERENCES

1) Bouma J., Jongerius A., Boersma O., Jager A. and Schoonderbeek P. 1977. The function of different types of macropores during saturated flow through four swelling soil horizons. Soil Sci. Soc. Am. J., 41, 945-950.

2) Greenland D.J. 1977. Soil damage by intensive arable cultivation: temporary or permanent? Phil. trans. R. Soc. London, 281, 193-208.

3) Ismail S.N.A. 1975. Micromorphometric soil porosity characterization by means of electro-optical image analysis (Quantimet 720). Soil Survey paper N° 9, Netherland Soil Survey Institute, Wageningen, pp. 104.

4) Jongerius A. and Heintzberger G. 1975. Methods in soil micromorphology. A technique for the preparation of large thin sections. Soil Survey paper N° 10, Netherland Soil Survey Institute, Wageningen, pp. 48.

5) Murphy C.P., Bullock P. and Turner R.H. 1977. The measurement and characterization of voids in soil thin sections by image analysis. Part I. Principles and techniques. J. Soil Sci., 28, 498-508.

6) Pagliai M. 1982. Influenza della somministrazione di liquami zootecnici sulle caratteristiche fisiche del terreno. In: "Somministrazione al terreno degli effluenti di allevamenti intensivi zootecnici". Collana del Progetto Finalizzato "Promozione della Qualità dell'Ambiente", CNR AC/4 121-133, 67-79.

7) Pagliai M., Guidi G. and La Marca M. 1980. Macro and micromorphometric investigation on soil dextran interactions. J. Soil Sci., 31, 493-504.

8) Pagliai M., Guidi G., La Marca M., Giachetti M. and Lucamante G. 1981. Effect of sewage sludges and composts on soil porosity and aggregation. J. Environ. Qual., 10, 556-561.

9) Pagliai M., La Marca M. and Lucamante G. 1983. Micromorphometric and micromorphological investigations of a clay soil in viticulture under zero and conventional tillage. J. Soil Sci., 34, 391-403.

10) Pagliai M. and Sequi P. 1981. The influence of applications of slurries on soil properties related to run-off. Experimental approach in Italy. In: J.C. Brogan (ed.) Nitrogen losses and surface run-off from landspreading of manures. Martinus Nijhoff/Dr. W. Junk Publishers, London, p. 44-65.

11) Tester C.F., Sikora C.J., Taylor J.H. and Parr J.F. 1981. Decomposition of sewage sludge compost in soil: 1. Carbon and Nitrogen Transformation. J. Environ. Qual., 6, 459-463.

RESULTS OF LARGE-SCALE FIELD EXPERIMENTS WITH SEWAGE SLUDGE AS AN ORGANIC
FERTILIZER FOR ARABLE SOILS IN DIFFERENT REGIONS OF THE NETHERLANDS

S. DE HAAN
Institute for Soil Fertility, Haren (Gr.), The Netherlands

Summary

In 1977 a series of six field experiments was started in which sewage
sludge was applied to arable soils in amounts of 0, 10 and 20 t DM/ha
to root crops (sugar beet and potatoes), normally grown every second
year, but sometimes more frequently, alternating with cereals. In the
years of sludge application large plots (30 m x 30 m) were subdivided
into five small plots receiving different rates of mineral fertilizer
N so as to make it possible to construct yield curves with which the
nitrogen effect of the sludge could be measured and maximum yields
determined without and with sewage sludge. In 1982 the effect of
sludge application on soil characteristics was determined. Results up
to 1982 are dealt with in this paper.
Sludge was found to be a rather unpredictable source of nitrogen. Maximum yield of potatoes was increased by sludge application in a number
of cases, and there was a positive after-effect on grain yield of cereals. The phosphorus effect of sludge was apparent mainly from a considerable improvement of the phosphorus status of the soils. Following
sludge application, concentrations of microelements increased in soils
in general, but not to an alarming degree, and in crops, in so far as
determined, only insignificantly.

1. INTRODUCTION

On the initiative and with sponsoring of the Foundation for Applied
Research of Waste Water in Rijswijk, The Netherlands, a series of six
large-scale field experiments was started in 1977 to study the positive
effects of sewage sludge, mainly due to its N, P and OM content, and its
possible negative effects, due to its heavy-metal content. The most interesting effect of sewage sludge, like that of organic fertilizers in general, may be the so-called rest-effect, which is an effect on yield that
cannot be achieved with mineral fertilizers alone applied in optimum
amounts. The rest-effect may be an "organic-matter effect" due to slow release of nutrients, or an effect on soil physical conditions, or of mineral elements in the organic fertilizer not accounted for in the treatments with mineral fertilizers alone. The rest-effect is positive in general, but may be negative in the case of sewage sludge due to an excess of
microelements. The research was carried out under supervision of the
Research Station for Arable Farming and Field Production of Vegetables in
Lelystad, in co-operation with the Institute for Soil Fertility in Haren.
Results obtained in the period 1977/82 are reported in this paper.

2. MATERIALS AND METHODS

The experiments were carried out on a reclaimed peat soil (AGM 316),
a loess (WR 158), and loams of marine origin (BEM 265, FH 86, KL 289 and

RH 400). Some of the most relevant characteristics of the soils at the beginning of the experiments in 1977 are presented in table I. The soils were analyzed again in 1982 for the same characteristics as in 1977 and every year for mineral N in the profile (0-20, 20-40 and 40-100 cm) at the end of the winter period, before crop growth started (Nmin).

Table I. Soil characteristics at the beginning of the experiments

Soil characteristic	Experimental field					
	AGM 316	BEM 265	FH 86	KL 289	RH 400	WR 158
pH-KCl (1n)	5.8	7.2	7.1	7.1	7.3	6.3
< 16 µm, %	-	26	12	52	40	28
Org. matter, %	14.7	3.9	2.0	4.0	5.0	3.2
N (total), %	0.24	0.11	0.07	0.18	0.13	0.10
P-water, mg/l	25	10	57	20	8	29
K-HCl, mg/100 g	11	16	14	50	27	14
CaO, %	0.86	6.72	0.66	5.74	8.28	0.35
MgO/NaCl, mg/kg	134	74	133	336	138	111
Cd (tot.), mg/kg	0.18	0.07	0.04	0.05	0.06	0.46
Cu, mg/kg	18.1	24.2	5.3	17.0	9.9	13.7
Mn, "	216	500	178	854	384	483
Ni, "	1.8	14.1	7.0	27.9	17.6	12.3
Zn, "	27	56	29	100	64	87

The crops grown in 1977-82 are shown in table II. The normal rotation was sugar beet - winter wheat - potatoes - winter wheat, but there are exceptions. In the six-year period root crops were grown four times on WR 158 and five times on AGM 316.

Table II. Crops grown on the experimental fields in the period 1977/1982.

Year	Experimental field					
	AGM 316	BEM 265	FH 86	KL 289	RH 400	WR 158
1977	s.b.	s.b.	s.b.	pot.	s.b.	s.b.
1978	pot.	w.w.	s.w.	w.w.	w.w.	pot.
1979	s.w.	pot.	pot.	s.b.	pot.	w.b.
1980	pot.	w.w.	w.w.	oats	w.w.	s.b.
1981	s.b.	s.b.	s.b.	pot.	s.b.	w.w.
1982	pot.	w.w.	w.w.	w.w.	w.w.	pot.

s.b.: sugar beet; pot.: potatoes; s.w.: spring wheat; w.w.: winter wheat; w.b.: winter barley

Where sludge was applied to root crops, it normally was applied three times, but on WR 158 four times and on AGM 316 five times. The rates of application were 0, 10 and 20 t DM/ha each time. Before application the sludges had been dewatered on drying beds, except the first three times in AGM 316, when liquid sludge was applied. Main sludge characteristics, averaged over the 1977/82 period, are presented in table III. It is well known that sludge composition is rather variable. N content, especially mineral N content, is normally much higher in liquid than in dried sludge, and this is reflected in the rather high N content of the sludge in the case of AGM 316. Heavy-metal contents were all well below the limits mentioned in the Dutch Guideline for the use of liquid digested sludge on arable and grassland (As, Cd, Hg 10; Ni 100; Cr, Pb 500; Cu 600 and Zn 2000 mg/kg DM). Sludges were applied in autumn before the root crop was grown, except in AGM 316 where application was in spring.

Table III. Average characteristics of the sewage sludges used on the experimental fields in 1977-1982.

Characteristic	Experimental field					
	AGM 316	BEM 265	FH 86	KL 289	RH 400	WR 158
pH (H$_2$O)	7.2	7.3	6.8	7.2	7.4	7.1
Org. mat., %	47.7	37.4	53.5	38.3	64.5	61.6
N (tot.), %	4.83	2.66	2.11	1.75	3.00	2.20
N (min.), %	1.82	0.84	0.19	0.44	0.25	0.18
P$_2$O$_5$, %	3.35	4.82	2.81	1.83	3.76	4.11
K$_2$O, %	0.46	0.24	0.24	0.13	0.27	0.22
CaO, %	2.36	4.46	3.15	1.82	4.07	3.70
MgO, %	0.42	0.53	0.39	0.19	0.33	0.35
Na$_2$O, %	0.25	0.11	0.10	0.06	0.11	0.05
Cl, %	0.24	0.12	0.12	0.05	0.16	0.06
Cd, mg/kg	4.33	4.77	3.75	2.78	4.99	5.66
Cu, "	533	308	400	82	480	287
Mn, "	199	1556	394	574	542	700
Ni, "	17	25	15	24	29	84
Zn, "	930	1588	868	536	1193	1071

Superimposed on the three sludge application rates were five mineral N application rates for root crops, making it possible to construct yield curves to measure the nitrogen effect of the sludges (generally their main nutrient effect), and to assess whether or not maximum yields were increased, or decreased, by sludge application.

The mineral N application rates were 0, 50, 100, 150 and 200 kg N/ha. No allowance was made for sludge nutrients. Mineral fertilizers, except N, were applied in optimum amounts on the basis of the zero sludge treatment. With cereals, sometimes allowance was made for differences in the amounts of Nmin in the profile due to the presence of sludge. Root crops were analyzed per sludge rate and per nitrogen rate for contents of nitrogen and phosphorus. In addition, sugar beet were analyzed for contents of sugar, and of α-amino N, K and Na in the sugar filtrate, being factors affecting the extractability of the sugar. In potatoes, in most cases underwater weight (a measure of the dry-matter and starch contents) and size grades (< 35, 35-55, > 55 mm) were determined. All crops were analyzed, per sludge rate and the highest N application rate only, for Cd, Cu, Mn, Ni and Zn, being the heavy metals generally most strongly affected by sludge application. In the case of manganese, the effect of sludge may be negative as a consequence of Zn-Mn-antagonism and/or a negative effect of lime in the sludge on Mn availability.

3. RESULTS

3.1. Nmin in the soil profile at the end of the winter period

The amount of fertilizer N to be applied to the crop depends on the amount of Nmin in the soil profile at the end of the winter period, which is very variable, especially because of variation in the precipitation surplus in the winter period. For this reason Nmin is determined at the end of the winter period in the soil profile to depth which has been fixed now at 60 cm for clay soils and 30 cm for potatoes on peat and sandy soils in The Netherlands, but which was still 100 cm at the start of the experiments.

The amounts of Nmin found in our experiments for the soils without sludge (SO) and the increase due to sludge, averaged over the two application rates ((S1+S2)/2 − SO), are presented in figure 1. Nmin was not determined in AGM 316 and RH 400 in 1977 and in KL 289 in 1979.

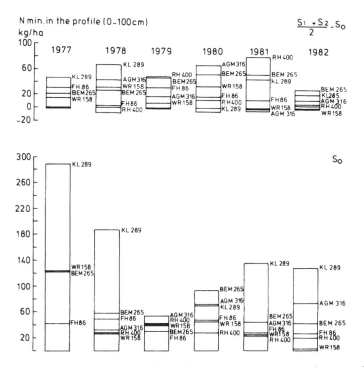

Figure 1. Mineral N in the soil profile at the end of the winter period without sewage sludge (SO) and increase due to sewage sludge, averaged over the sludge application rates ((S1+S2)/2 − SO; SO, S1, S2 = 0, 10 and 20 t sludge DM/ha to root crops).

There appears to be considerable variation in Nmin between fields and years, but without any relationship to the precipitation in the winter period (November to February). The average amount was 64 kg per ha per year for SO, and the increase due to sewage sludge was 17 kg N per 10 t sludge

DM. This increase was due partly to Nmin in the sludge, in so far as it was applied in autumn, and partly to mineralization of sludge organic N. In the years without sludge application, when cereals were grown, the sludge organic N is the main source. In these years the average value for Nmin was 10 kg/10 t sludge DM. In the years when root crops were grown the value was 24 kg N/ha.

3.2. Results with sugar beet

3.2.1. Beet yield

Curves for net beet yield without sewage sludge and increases due to sewage sludge are presented in figure 2 for the individual years and for average values over the years.

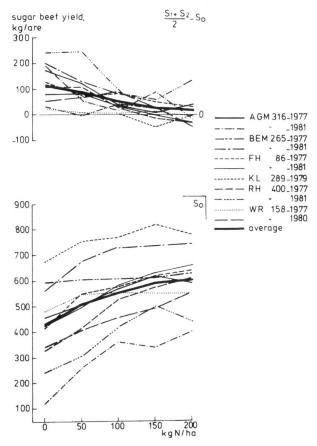

Figure 2. Sugar beet yield without sewage sludge (SO) and increase due to sewage sludge, averaged over the sludge application rates ((S1+S2)/2 - SO; SO, S1, S2:cf. fig. 1).

There was much variation between soils and between years. This was especially the case when no mineral fertilizer N and no sewage sludge was applied; the explanation could be the variation in Nmin in the soil profile at the end of the winter period. However, the correlation between the two variables was statistically not significant (r = 0.57). The same is true for the increase in yield and the increase in Nmin due to sewage sludge (r = 0.54).

Yield increase due to sewage sludge generally decreased with increasing mineral fertilizer N application rate, becoming negative in a number of cases at the highest N application rates, due to excess nitrogen.

3.2.2. Nitrogen effect of soils and sludges, partitioned into Nmin in the soil profile at the end of the winter period and Nmin released from organic N during the growing season

The nitrogen effect of sludges as well as of soils can be calculated from beet yield as indicated in figure 3 for FH 86-1977. In this figure yields are expressed as values of the equation $y = ax^2 + bx + c$ (I), in which y stands for yield and x for N application rate. The nitrogen effect of the soil is the extrapolated absolute value of x at y = 0, for sludge it is the value of x at $y = c1$ or $c2$ (for S1 and S2, respectively).

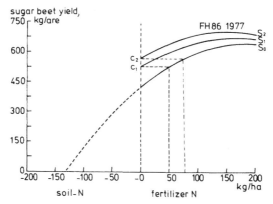

Figure 3. Determination of soil (S0) and sludge (S1, S2) nitrogen effect from sugar beet yield of FH 86-1977 calculated with the equation $y = ax^2 + bx + c$, in which y = yield and x = N application rate.

By subtracting Nmin in the soil profile at the end of the winter period from the total nitrogen effect, a value is obtained for mineral N released from soil or sludge OM during the growing season. The result of this calculation is presented in table IV. It is apparent that soil as well as sludge OM are rather unpredictable N suppliers. There is no relationship with soil or sludge organic N. The mean value for N released from sludge OM (15 t sludge DM) was 56 kg/ha, or about 15% of sludge organic N. The corresponding value for soil OM was 104 kg/ha or about 3% of soil organic N in the 20 cm top layer. However, soil N may partly originate from deeper layers.

Years in which Nmin was not determined have been left out of consideration in table IV, and it was assumed that no loss of N occurred during

Table IV. Soil (SO) and sludge((S1+S2)/2 -SO) nitrogen effect in kg N/ha (a), calculated from beet yield,and its partitioning into Nmin in the soil profile at the end of the winter period (b) and Nmin released from soil or sludge organic matter during the growing season (c).

	AGM 316 1981	BEM 265 1977	1981	FH 86 1977	1981	RH 400 1981	WR 158 1977	1980
				SO				
a	39	281	162	132	160	68	237	224
b	43	121	44	42	27	23	123	46
c	-4	160	118	90	133	45	114	178
				(S1+S2)/2 -SO				
a	73	139	180	62	50	156	22	40
b	5	48	54	30	14	81	16	34
c	68	91	126	32	36	75	6	16

the growing season.

3.2.3. Concentrations of sugar and nitrogen in the beet and of α-amino N, K and Na in the sugar filtrate

These concentrations are presented in table V for SO and (S1+S2)/2, averaged over the years in which sugar beet was grown.

The effect of mineral fertilizer and of sludge on the sugar concentration was negative; the effect of sludge was rather strong compared with its effect on beet yield. The nitrogen concentration was increased by fertilizer N as well as by sewage sludge. The correlation between nitrogen and sugar concentrations was statistically highly significant (r = -0.71****), but with considerable variation between years.

Table V. Contents of sugar (% of fresh matter) and N (% DM) in beet and of α-amino N, K and Na (meq/100 g sugar) in sugar filtrate without (SO) and with sewage sludge ((S1+S2)/2), averaged over the experiments.

		N applied, kg/ha				
		0	50	100	150	200
Sugar	SO	17.22	17.27	17.12	16.93	16.59
	(S1+S2)/2	16.52	16.50	16.26	16.07	15.75
N	SO	0.62	0.63	0.67	0.72	0.77
	(S1+S2)/2	0.70	0.74	0.78	0.82	0.86
α-am.	SO	10.8	11.3	12.6	14.5	17.2
	(S1+S2)/2	15.8	16.9	19.3	20.9	23.4
K	SO	29.7	29.0	28.7	29.4	30.3
	(S1+S2)/2	31.9	31.4	32.1	32.6	34.0
Na	SO	2.0	2.3	2.5	2.8	3.2
	(S1+S2)/2	3.5	3.6	4.3	4.5	5.0

A high N concentration in beets may be accompanied by a high concentration of α-amino N in the sugar filtrate which has to be neutralized by addition of lime, which lowers the amount of extractable sugar. The critical value is 17 meq α-amino N/100 g sugar. Each additional unit lowers the amount of extractable sugar by about 0.5%. The value for α-amino N was increased by mineral fertilizer N as well as by sewage sludge, but especially by the latter. Most likely it was increased especially by N released from OM towards the end of the growing season. In BEM 265, KL 289 and WR 158 there was a good correlation between N in beet and α-amino N in the sugar filtrate (r = 0.95[****]), in the others there was no relationship (r = - 0.09).

K and Na also have a negative effect on sugar extractability. The decrease in amount of extractable sugar is about 0.3% per unit of K or Na, expressed in meq/100 g sugar. Concentrations of both K and Na were increased by sewage sludge, but not, or only slightly, by mineral fertilizer N. The effect of sewage sludge may be due to K and Na in the sludges, which was not taken into consideration in the mineral fertilization.

3.2.4. Sugar yield

Sugar yield is the most important yield characteristic, as farmers are paid for sugar, not for beet yield, and so far not for extractable sugar.

The picture for sugar yield is about the same as that for beet yield, but with one important difference: at higher N application rates, the sludge effect became negative in many more cases than with beet yield, due to the negative effect of sludge on beet sugar concentration. The mean sludge effect became negative at an N application rate of about 140 kg/ha (cf. figure 5).

3.2.5. Maximum sugar yields and optimum N application rates

Maximum sugar yields and corresponding optimum mineral fertilizer N application rates can be calculated easily as indicated in figure 4 for FH 86-1977. In this figure yields are expressed again as values of the equation $y = ax^2 + bx + c$ (I). Maximum yields here are economically maximum yields, obtained together with the corresponding N application rates by

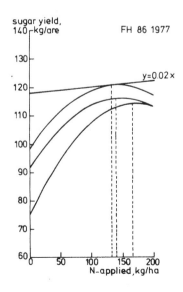

Figure 4. Determination of economically maximum yields and corresponding optimum N application rates as coordinates of the contact points of y = 0.02 x (0.02 = price kg N/price 100 kg sugar = Hfl 1.50/Hfl 75.-) and $y = ax^2 + bx + c$ (cf. fig. 3).

calculating the coordinates of the contact points of eq. (I) with $y = px + q$, in which p is the ratio between the price per unit of N (Dfl. 1.50/kg) and the price per unit of product (Dfl. 75/100 kg sugar).

For the various fields and years, maximum yields and corresponding optimum N application rates, and in- or decreases due to sewage sludge, averaged over the two application rates, are presented in table VI. The mean maximum sugar yield was 102 kg/are, with a variation from 65 to 133 kg/are, for which no explanation can be given. The mean increase due to sewage sludge was zero with a rather small variation, and none of the individual values was statistically significant.

Table VI. Economically maximum sugar yields in kg/are (a) and corresponding optimum N application rates in kg/ha (b) without sewage sludge (SO) and increases due to sewage sludge, averaged over sludge application rates $((S1+S2)/2 - SO)$.

	AGM 316		BEM 265		FH 86		KL 289	RH 400		WR 158		AVERAGE
	1977	1981	1977	1981	1977	1981	1979	1977	1981	1977	1981	
					SO							
a	106	66	107	128	114	113	133	99	80	88	88	102
b	140	150	0	130	170	200	120	180	160	100	200	140
					$((S1+S2)/2 - SO)$							
a	-2	-1	-2	5	4	-4	-4	-2	8	-2	-2	0
b	-140	-20	0	-80	-30	-60	-10	-30	-100	0	-90	-50

Without sewage sludge the mean optimum N rate was 140 kg/ha, with 0 and 200 kg/ha as assumed extreme values and 70% of the variation accounted for by Nmin in the soil profile at the end of the winter period ($r = -0.83^3$). The mean decrease due to sewage sludge was 50 kg/ha, with 0 and 140 kg/ha as extreme values and only 30% of the decrease accounted for by the increase in Nmin in the soil profile due to sewage sludge ($r = -0.56$).

3.2.6. Beet phosphorus concentration

There was no effect of mineral fertilizer N on beet P concentration and only a small effect of sewage sludge. The mean concentrations were 0.40 and 0.42% P_2O_5 in DM without and with sewage sludge, respectively.

3.2.7. Beet heavy-metal concentrations

Concentrations of Cd, Cu, Mn, Ni and Zn are presented in table VII for each sludge application rate, averaged over the years in which sugar beet was grown. The variation between years and soils was rather large. Only the zinc concentration was increased significantly (P > 99%) due to sewage sludge.

3.2.8. N, P, Cd, Cu, Mn, Ni and Zn in sugar beet leaves + heads

In general, concentrations in leaves + heads were about three times as high as in the beet, but for P this value was much lower. If mean concentrations in beet DM are set at 100, then values for leaves + heads are as follows: N 272, P 174, Cd 313, Cu 257, Mn 279, Ni 310 and Zn 302. The r-values for the correlation between concentrations in beet and in leaves + heads were: N 0.79[****], P 0.56[****], Cd 0.86[****], Cu 0.47[***], Mn 0.89[****], Ni -0.19 and Zn 0.90[****].

Table VII. Concentrations of Cd, Cu, Mn, Ni and Zn in sugar beet in mg/kg
DM, averaged over 11 years in which sugar beet was grown. S0, S1, S2:
sludge application rates of 0, 10 and 20 t DM/ha to root crops.

	Cd	Cu	Mn	Ni	Zn
S0	0.17	3.9	37	0.3	22
S1	0.18	4.0	41	0.4	29
S2	0.19	4.0	36	0.3	29

3.2.9. Main effects of sewage sludge on sugar beet

The main effects of sewage sludge on sugar beet are summarized in
figure 5 in which yields of sugar and of beet and amounts of N in beet and
in leaves + heads, and of α-amino N, K and Na in the sugar filtrate, with
sewage sludge, averaged over the two application rates, are expressed as a
percentage of the corresponding values without sewage sludge, which were
set at 100 for every mineral fertilizer N application rate.

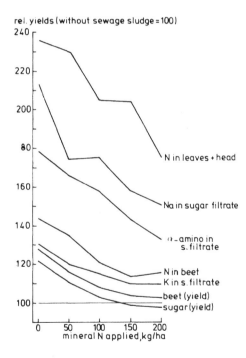

Figure 5. Yields (amounts)
with sewage sludge, averaged
over the sludge application
rates and years in which sugar
beet were grown, in % of
corresponding values without
sewage sludge, which for every
N application rate were set
at 100.

Sludge effects decreased with increasing mineral fertilizer N application rate. The effect on sugar yield became negative at an N application
rate of 140 kg/ha, which was the optimum N application rate without sewage
sludge. With sewage sludge the optimum N application rate was 90 kg/ha.
According to figure 5, at this N application rate there was a small positive rest-effect of sewage sludge on sugar yield (5%), and a larger one on

beet yield (10%), amount of N in beet (23%) and leaves + heads (110%), and amount of α-amino N (60%), K (16%) and Na (73%) in the sugar filtrate. Positive effects on the last three constitute, as we know, negative effects on the amount of extractable sugar.

Sewage sludge apparently increases primarily the amounts of nitrogen in leaves + head and beet, and of α-amino N, K and Na in the sugar filtrate, and to a much lesser degree beet and sugar yield; the net result is that the amount of extractable sugar at the optimum mineral fertilizer N application rate is not affected or even reduced.

3.3. Results with potatoes

3.3.1. Tuber yield

The effect of mineral fertilizer N and sewage sludge on tuber yield is presented in figure 6. Without sewage sludge the lowest yield was obtained on FH 86-1977, due to the fact that seed potatoes were grown here, which are harvested in an early stage. The variety grown was Bintje in all cases, except on AGM 316, which lies in a region where potatoes are grown for starch production.

The effect of sewage sludge decreased with increasing rate of mineral fertilizer N. However, the effect remained positive up to the highest N application rate in almost all cases.

3.3.2. Nitrogen effect of soils and sludges, calculated from tuber yield, partitioned into Nmin in the soil profile at the end of the winter period. and N released from soil or sludge OM during the growing season

These effects were calculated in the same way as for sugar beet. Again, soil and sludge OM turned out to be rather unpredictable N suppliers. N released from soil OM varied from 0 - 200 kg/ha with 80 kg/ha as an average. For sludge OM the average was about 50 kg/ha with 0 and 100 kg/ha as extreme values. No exact values could be calculated for AGM 316 as yields with sewage sludge alone were already higher than maximum yields without sludge.

3.3.3. Size grades

The largest grade (> 55 mm) was most strongly affected by sewage sludge application. Even at the highest N application rate, yield of this grade with sewage sludge, averaged over the application rates and years, was more than 120% of the yield without sewage sludge (cf. fig. 7). The other size grades (< 35 and 35 - 55 mm) were much less affected. At the highest N application rate the sludge effect was slightly negative here.

3.3.4. Maximum tuber yields and optimum N application rates

Maximum tuber yields and corresponding optimum mineral N application rates were calculated as reported for sugar beet, assuming a price of Dfl. 0.30/kg for ware potatoes and of Dfl. 0.15 for potatoes for starch production in the case of AGM 316. In this case yield is measured as "weight for payment", which is tuber yield multiplied by a factor K = (underwater weight - 100)/300. The result of the calculation is presented in table VIII.

Maximum tuber yields varied from about 460 to 620 kg/are with 520 kg/are as an average. There was a significant increase in maximum yield due to sewage sludge in a number of cases (AGM 316-1978 and 1980, BEM 265, RH 400), in other cases increases were negligible or even negative (FH 86). The mean increase was 25 kg/are.

Optimum mineral N application rates without sewage sludge varied from 110 to an assumed maximum of 300 kg/ha with 190 kg/ha as the average value.

68

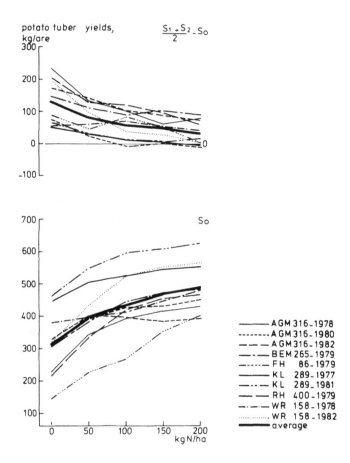

Figure 6. Potato tuber yields without sewage sludge (SO) and increases due
to sewage sludge, averaged over the sludge application rates ((S1+S2)/2
- SO; SO, S1, S2: cf. fig. 1).

There was no significant correlation with Nmin in the soil profile at the
end of the winter period (r = -0.19). Decreases in optimum N application
rate due to sewage sludge varied from 110 to -90 kg/ha with 20 kg/ha as the
average value and without a significant correlation with the increase in
Nmin in the soil profile at the end of the winter period (r = 0.34).

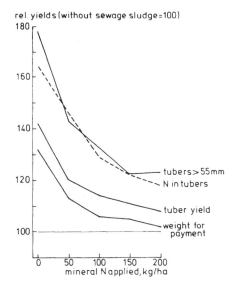

rel. yields (without sewage sludge=100)

- tubers > 55mm
- N in tubers
- tuber yield
- weight for payment

mineral N applied, kg/ha

Figure 7. Yields with sewage sludge, averaged over the sludge application rates and years in which potatoes were grown, in % of corresponding values without sewage sludge, which for every N application rate were set at 100.

Table VIII. Economically maximum tuber yields in kg/are (a) and corresponding optimum N application rates in kg/ha (b) without sewage sludge (S0) and increases due to sewage sludge, averaged over sludge application rates ((S1+S2)/2 - S0).

	AGM 316			BEM 265	FH 86	KL 289		RH 400	WR 158		AVERAGE
	1978	1980	1982	1979	1979	1977	1981	1979	1978	1982	
						S0					
a	463	509	470	487	517	554	624	470	560	566	520
b	150	150	110	180	300	190	180	180	300	180	190
						((S1+S2)/2 - S0)					
a	38	75	10	100	-78	0	33	80	-15	14	25
b	-100	-25	-110	35	-30	-10	90	-30	-35	25	-20

3.3.5. Underwater weight and nitrogen and phosphorus contents of tubers

These characteristics, without and with sewage sludge, averaged over the sludge application rates and years, are presented in table IX.

Like the sugar content of beet, the underwater weight of potatoes decreased with increasing N application rate. The sludge effect was negative, in contrast with the effect on the nitrogen content. There was a rather close correlation between these two characteristics ($r = -0.82^{****}$). The phosphorus content of the tubers was slightly decreased by mineral fertilizer N (as a consequence of "dilution") and increased by sewage sludge. The level of $0.50\% \ P_2O_5$ in DM, necessary for optimum tuber growth, was not reached in some cases (KL 289-1977, RH 400-1979). In these cases the possibility of a positive effect of sludge phosphorus on tuber yield cannot be excluded.

Table IX. Underwater weight and contents of N and P_2O_5 (% of DM) of potato tubers without (SO) and with sewage sludge ((S1+S2)/2), averaged over the years in which potatoes were grown.

		N application rate, kg/ha				
		0	50	100	150	200
Underwater	SO	447	440	431	418	410
weight	(S1+S2)/2	424	420	407	404	395
N	SO	1.08	1.13	1.31	1.43	1.55
	(S1+S2)/2	1.34	1.43	1.56	1.66	1.76
P	SO	0.54	0.50	0.48	0.49	0.47
	(S1+S2)/2	0.56	0.54	0.54	0.53	0.54

3.3.6. Cd, Cu, Mn, Ni and Zn in tubers

Concentrations of these elements are presented in table X averaged over ten years in which potatoes were grown. Only copper and zinc were significantly (P > 95%) increased by sewage sludge. The critical Cd level of 0.4 mg/kg DM for ware potatoes was exceeded in a single case (WR 158-1978) due to sewage sludge.

Table X. Concentrations of Cd, Cu, Mn, Ni and Zn in potato tubers in mg/kg DM, averaged over 10 years in which potatoes were grown. SO, S1 and S2 = sludge application rates of 0, 10 and 20 t DM/ha to root crops.

	Cd	Cu	Mn	Ni	Zn
SO	0.12	4.0	5	0.3	13
S1	0.14	4.7	5	0.3	15
S2	0.14	4.8	5	0.3	15

3.3.7. Main effects of sewage sludge on potatoes

In figure 7 total tuber yield, yield of tubers > 55 mm, weight for payment and N in tubers in kg/ha, averaged over the two sludge application rates and the years in which potatoes were grown, are expressed as a percentage of the corresponding values without sewage sludge, which again were set at 100 for every N application rate. The figure shows that also in the case of potatoes the sludge effect was primarily a nitrogen effect, which decreased in the order N in tubers = yield of tubers > 55 mm > tuber yield > weight for payment.

3.4. Results with cereals

3.4.1. Yield

With only one exception (WR 158-1979; winter barley), there was a positive after-effect of sludge application on grain yield. Without sewage sludge the average yield was 7120 kg/ha (85% DM), and with sewage sludge 7640 kg/ha. Most probably the increase in yield was primarily an effect of N released from sludge OM during the growing season. In some cases less mineral fertilizer N was applied with sewage sludge, because of an increase in Nmin in the soil profile at the end of the winter period. The mean increase was 1 kg N/t sludge DM. The mean decrease in fertilizer N was 8 kg/ha.

3.4.2. Heavy metals in grains

Concentrations of Cd, Cu, Mn, Ni and Zn in grains, averaged over 14 years in which they were determined (not in WR 158-1979) are presented in table XI. Only the zinc concentration was increased significantly due to sewage sludge application.

Table XI. Concentrations of Cd, Cu, Mn, Ni and Zn in grain in mg/kg DM averaged over 14 years in which cereals were grown. S0, S1, S2 = 0, 10 and 20 t sludge DM/ha, applied to root crops.

	Cd	Cu	Mn	Ni	Zn
S0	0.10	4.0	34	0.3	32
S1	0.12	4.1	34	0.2	33
S2	0.12	4.1	35	0.2	35

3.4.3. Results of soil analysis in 1982

When values for the treatments without sewage sludge, averaged over the experimental fields, are set at 100, then the corresponding values for treatments with sludge, averaged over the two application rates, are for pH-KCl 100, humus 113, N(total) 115, P(total) 122, P(water) 173, K(0.1 N HCl) 98, Ca(total) 100, Mg(total 100, MgO/NaCl 111, Na(total) 114, Cl 91, Cd(total) 120, Cu(total) 162, Mn(total) 101, Ni(total) 103 and Zn(total) 126.

Concentrations of especially water-soluble P and total Cu were increased by sewage sludge. All heavy-metal concentrations remained well below critical limits. Of most practical importance is the improvement of the phosphorus status of the soils. Values for water-soluble P (Pw-values) for individual fields and sludge application rates are presented in table XII.

Table XII. Pw-values (mg water-soluble P_2O_5/l soil) in 1982 as affected by sewage sludge application in the period 1977/1982.

Exp. field		AGM 316	BEM 265	FH 86	KL 289	RH 400	WR 158
ss application		5 x	3 x	3 x	3 x	3 x	4 x
t DM/ha/appl.	0	45	13	46	24	18	31
	10	98	18	52	26	37	36
	20	118	27	66	29	50	55

4. Conclusions

From the experiments the following conclusions can be drawn:

1. The main effect of sewage sludge is a nitrogen effect, which cannot be predicted exactly. When sludge is applied in autumn, mineral nitrogen in the soil profile at the end of the winter period should be determined as a measure of the sludge nitrogen effect to be expected for the next growing season. When sludge is applied in spring, only the mineral nitrogen in the sludge should be taken into account.

2. In a rotation consisting of sugar beet, potatoes and cereals, sludge should be preferably be applied to potatoes, as it is a crop which

may respond to nitrogen released from soil or sludge organic matter during the growing season with higher yields than can be obtained with mineral fertilizers alone. With sugar beet, the maximum beet yield may be somewhat higher with sewage sludge, but not the maximum sugar yield because of an adverse sludge effect on the sugar content. Moreover, sugar extractability may be adversely affected by sewage sludge.

3. A positive after-effect of sludge application may be expected as was demonstrated in these experiments by increased grain yields of cereals, most probably due to an improved nitrogen status of the soils.

4. The phosphorus effect of sewage sludge manifested itself in these experiments especially in a considerable improvement of the phosphorus status of the soils.

5. Application of sewage sludge resulted in an enrichment of the soils with microelements, most of them heavy metals, which can be regarded as positive in the case of copper, zinc and boron, but negative in the case of elements like arsenic, cadmium, mercury and lead. All microelement levels remained well below safe limits for soils and the enrichment was not, or only slightly, reflected in the heavy-metal concentrations determined in the crops.

THE CUMULATIVE AND RESIDUAL EFFECTS OF SEWAGE SLUDGE NITROGEN ON CROP GROWTH

J E HALL
Water Research Centre, Medmenham Laboratory, UK.

Summary

Sewage sludges contain nitrogen in readily available inorganic and slow release organic forms. Organically-bound nitrogen is mineralised slowly over an extended period of time and its availability is governed by the degree of sludge stabilisation and its C:N ratio as well as soil and climatic factors. Results from a number of experiments are described involving single and repeated applications of different types of sludge to grassland which show effects within and between years attributable to the residual sludge organic N. Nitrogen is released from a single sludge application in a predictable long term manner with predictable accumulated residual effects following regular applications of sludge. This information would allow reductions in the amounts of additional N fertiliser required where sludge is applied either on a regular basis to farmland or as large one-off applications for land reclamation.

1. INTRODUCTION

The use of sewage sludge as a nitrogen source is widely recognised, its short term fertilising effects are now quite well understood and much of the published data have been reviewed along with practical advice for sewage treatment authorities and farmers (1,2,3).
Nitrogen occurs in sludge in ammoniacal and organic forms. Whilst NH_4-N is immediately available for crop uptake, organic N requires mineralisation first, the rate of which depends on the C:N ratio of the sludge as well as the degree of stabilisation of the organic matter. Table 1 summarises the amounts of organic N found to be available in the year of application from different types of sludge (2). The remaining organic matter will contain a significant proportion of the sludge nitrogen originally applied but this will be much more resistant to mineralisation and consequently its value in the following year will be reduced. However there is very little data available on the likely magnitude of this residual value and of its rate of release in subsequent years. This information would be particularly valuable for land which receives sludge regularly where the accumulated residual effects could make a significant contribution in further reducing the requirement for nitrogen fertiliser.
This paper illustrates some of the cumulative and residual effects of sewage sludge derived from data drawn from a number of field trials instigated by the Water Research Centre; the main effects of many of these trials have already been reported (4-9 for example).

Table 1. Crop availability of sludge organic N in year of
          application (2)

Sludge type	Organic N available (%)	
Biological	50-60	
Primary	30-40	Increasing stability
Digested	15-20	of organic matter
Stored Digested	10	

2. EFFECTS ON THE ANNUAL GROWTH PATTERN OF GRASS

2.1 Single applications of sludge

It is well known that NH_4-N in sludge is as available for crop
uptake as nitrogen fertiliser (10) and that equivalent yields of grass
are produced where liquid digested sludge and N fertiliser are applied
at equivalent rates of inorganic N at the optimum time in short term
trials (11). Further trials, where three harvests were taken over the
growing season and where liquid digested sludge and fertiliser were
applied separately on four separate occasions over the winter period
at equivalent inorganic N rates, show that the organic N in the sludge
made a significant contribution particularly after the first cut of
grass when all the inorganic N was removed. This is illustrated by
Figure 1 which is drawn from the mean data of six separate
experiments (9).   The first cut of grass in the year showed
differences in yield attributable to the time of application; although
sludge nitrogen was apparently lost from the early applications,
particularly November, the losses from fertiliser were much greater as
might be expected. Whilst the best fertiliser response was from the
latest application in March, sludge produced the best initial yield
from the February application indicating that some of this nitrogen
had been derived from the organic nitrogen and that the March
application had been too late for sufficient mineralisation to take
place to improve on the response from the February application.

For both cuts 2 and 3 an additional 40 kg N/ha were applied to
sludge, fertiliser and a control treatment immediately after cuts 1
and 2.   Figure 1 shows that there was no residual response to
fertiliser applied over the winter at cuts 2 or 3 and by inference
that there would be no residual NH_4-N originally applied in the
sludge. However, the yield from sludge was higher than the control at
cut 2 and marginally so at cut 3 indicating the continued release of
organic N throughout the year from a single application of sludge.
Figure 1 also indicates that no organic N had been lost over the
winter as the grass yields at cut 2 and 3 were similar for sludge
applied at different times.

2.2 Repeated applications of sludge

As discussed later, repeated applications of sludge have a
cumulative effect on improving total annual grass yields, however
there are effects on the annual growth pattern as well. The normal
pattern in the UK is for rapid growth in the spring with limited
growth over the summer and a variable "autumn flush" depending on the

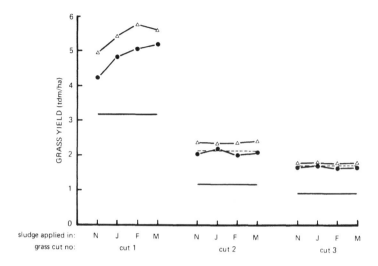

Figure 1. Yields of ryegrass from three successive cuts per year following separate applications of liquid digested sludge or fertiliser (applied at equivalent N rate to sludge inorganic N) in November (N), January (J), February (F) or March (M). Sludge (△), fertiliser (●) and a control (- - -) received 40 kg N/ha after both cut 1 and 2. Untreated control (———). These figures are the mean results from two sites over three years.

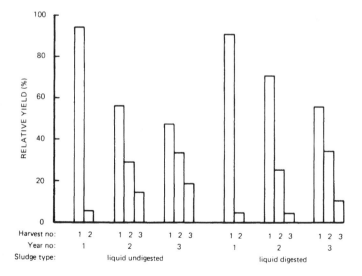

Figure 2. Changes in grass yield patterns following successive spring applications of liquid undigested and digested sludges. Figures are means of three soil types. Yield per harvest is as a percent of annual yield.

weather and nitrogen supply. As shown above, a single application of fertiliser at the start of the growing season has a negligible effect on growth after the first cut whereas mineralisation of sludge organic N will continue to supply the grass. Where sludge is applied annually, the contribution from organic N progressively increases such that the yields of cuts 2 and 3 in successive years increase relative to cut 1.

Figure 2 summarises the harvest data from trials on three different soil types (sandy, silt and clay loams) which received successive annual applications of liquid undigested (LUDS) and liquid stored digested sludges (LSDS) over three years. The data for each harvest is expressed as a per cent of the total yield for each year and sludge type. In the first year, virtually all the response was in the first harvest; this was a green barley crop undersown with ryegrass, the latter giving a low relative yield in cut 2 as the crop established. For both sludge types, the general growth pattern in subsequent years was similar. In the second year the initial cut was lower relative to that in the first year and similarly in the third year when the first cut was relatively lower than that in the second year. Both the second and third cuts of grass increased their relative yields in the third year, indicating an increasing pool of sludge organic N releasing nitrogen after the initial cut each year.

Liquid digested sludge contains more inorganic N than LUDS as is indicated by the higher relative yields for LSDS in Figure 2 for the first cuts in each year. However, the organic matter of LUDS, not being stabilised, would mineralise more rapidly than that of LSDS as is indicated by the higher relative yields for harvests 2 and 3 compared with those of LSDS.

The implication for the grassland farmer is that there will be improved late season growth where sludge is applied regularly when the response to fertiliser is often limited. Furthermore, it has been noticed by farmers that receive sludge regularly, grass growth often starts two to three weeks earlier in the spring indicating the carryover of nitrogen from one year being released early the next.

3. LONGER TERM EFFECTS

3.1 Single applications of sludge

The effect of a single application of liquid sludge beyond the year of application is generally very small due to the limited amount of organic matter, particularly in digested sludge, that is applied in an average dressing of sludge containing 2-5 tds/ha. Appreciable long-term effects from a single application are only observed with dewatered sludges where the addition of organic N may be quite large although the amount of nitrogen released will depend on whether the sludge was stabilised or not prior to dewatering.

Figure 3 summarises the results of two series of field trials on four soil types with dewatered undigested and air dried digested sludge. This shows the relative rate of release of nitrogen over a period of four years (4,8). Whilst the undigested sludge had the higher values in each year (e.g. 22% vs 13% in the first year) the decrease in N availability overtime was at the same rate for both sludge types in that the residual value in the second year was half of

the initial N availability, the value in the third year was half of
that in the second and similarly for the fourth year. This suggests
that the release of organically bound nitrogen is long term and
predictable with the residual N having a half-life of one year. This
observation is supported by the published data of others which is
summarised in Table 2.

Table 2.   Release of N(%) over time from single applications of
    sludge.

YEAR	de Haan* (12)	Keeney et al (13)	Kerr et al (14)
1	6.4	15	20
2	4.3	6	10
3	2.3	4	5
4	1.2	2	

*mean of 7 sludges

3.2 Repeated applications of sludge

    There is little published data available on the cumulative effects
of repeated applications of sludge.   Pratt et al in California, USA
proposed a decay series for a variety of organic wastes including an
unspecified liquid sewage sludge (15).   They suggested that 35% of
sludge organic N would be mineralised in the first year, 10% in the
second and 5% in all subsequent years with this series being applied
individually to each yearly application of sludge.   However, this
theoretical model has not been field tested and under more temperate
conditions the decay series would certainly be different.
    A series of 16 experiments instigated by the Water Research Centre
and conducted on five different soil types with LUDS and LSDS applied
annually in the spring or autumn for up to 5 years is summarised in
Table 3.   This shows the per cent cumulative recovery of sludge
nitrogen each year from the preceding sludge applications.   It is
clear that the variability of N recovery between each experiment is
large (5.0 to 27.4% in year 1) reflecting the problems of general
quantification in agricultural trials.   However, pooling the results
of 16 experiments should provide a reasonably accurate assessment of
the basic pattern of effects from repeated sludge applications
although no supporting statistical evidence is available yet.   Table 4
shows that sludge type and time of application has a relatively small
effect on the overall cumulative N recovery.   The mean N recovery of
LSDS was higher at 12.7% in the first year compared with 11.9% for
LUDS presumably due to its higher NH_4-N content.   However, in
subsequent years the mean recovery from LUDS is higher due to the
greater amounts of readily degradable organic matter applied.   Time of
application shows a small advantage to spring applied sludge in 3 out
of 4 years indicating that whilst there are winter losses from an
autumn application they are relatively small.

Table 3. Cumulative per cent recovery in ryegrass of nitrogen applied in successive annual applications of sludge (5, 6, 7)

S - Spring applied sludge
A - Autumn applied sludge

Sludge type:

	LIQUID UNDIGESTED SLUDGE								LIQUID STORED DIGESTED SLUDGE							
Soil type :	SAND		SILT		CLAY		LOAM		SAND		SILT		CLAY		CHALK	
Year of harvest	S	A	S	A	S	A	S	A	S	A	S	A	S	A	S	A
1978	5.0	–	15.4	–	6.4	–	–	–	6.0	–	9.7	–	8.2	–	–	–
1979	20.0	–	18.3	6.8	14.7	8.8	27.4	11.5	10.2	10.6	18.8	19.4	13.2	6.6	–	–
1980	21.6	13.8	20.0	13.6	15.9	18.8	36.8	18.8	15.2	14.0	22.5	23.6	16.6	11.2	21.8	19.2
1981	28.2	24.7	–	–	–	–	36.8	21.8	24.8	19.2	–	–	–	–	24.6	20.4
1982	34.2	24.4	–	–	–	–	(52.2)*	(35.7)*	29.0	24.2	–	–	–	–	–	–
	(34.7)*								(28.5)*							

* Cumulative recovery of sludge N in 1982 from previous applications where no additional sludge was applied in 1982.

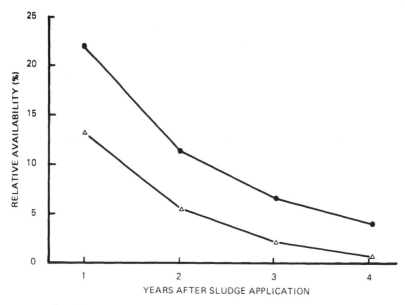

Figure 3. The rate of release of nitrogen from dewatered undigested (●) and digested (△) sludges relative to N fertiliser (4, 8).

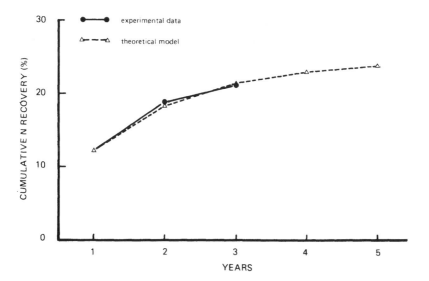

Figure 4. The cumulative recovery of nitrogen from successive annual applications of sludge.

The overall mean recovery of sludge nitrogen in the first year was 12.3%. If this figure is assumed to be an average value applicable to the initial recovery of sludge applied in any year, and that N release from a single application decreases by half each year as discussed above, a model to describe the cumulative residual values of regular sludge applications may be based on the geometric equation:

$$N \text{ recovery} = 2I \ [1-(1/2)^n]$$
in year n

where n = number of years after initial
        application
    I = initial N recovery at n=1

Table 5 and Figure 4 compare the mean cumulative N recovery data from Table 4 with that derived from the theoretical model. This shows very close agreement up to year 3 (the data for years 4 and 5 are limited to one site only and are not compared) and indicates that the greatest increase in residual value will be within three years of the initial application where sludge is applied annually and that additional benefits from residual sludge N will be relatively small thereafter.

The nitrogen value of LUDS in particular is reliant on rapid mineralisation for its full nitrogen fertiliser value in the year of application so when sludge is applied after the start of the growing season, or where adverse conditions for mineralisation follow application or if arable crops are grown, the first year response to LUDS may be poor. Table 6 summarises the results of a grassland trial (7) where LUDS was applied annually. It is clear that the first year yield response (relative to N fertiliser) was very poor at 7.5% whereas the relative nitrogen recovery was 28.8%. This indicates that mineralisation took place as sludge N had been absorbed but too late to stimulate growth. In the following year there was a large compensatory response with a relative yield of 76%. Weather records indicated difficult mineralisation conditions during the first year until early autumn. This suggests that the N released was absorbed by the turf mat during the autumn and utilised early in the following spring in dry matter production. Mean N efficiency was 40% after three years, however if year 4 is included when no sludge was applied the overall efficiency increased to 50% due to the accumulated residual sludge nitrogen.

The figures in Table 6 are based on grassland trials which may be expected to recover a much greater proportion of the mineralised nitrogen from a given application of sludge than arable crops where crop requirement for N does not generally coincide closely with peak mineralisation rates. Cereals in particular will not benefit from N released over the late summer/early autumn period when mineralisation is at its most rapid as such crops finish absorbing nitrogen in June. It has been shown that the sludge N value to cereals is about half of that to grass (7).

The results of winter wheat trials (16,17) summarised in Table 7 show similar effects as the grassland trials in Table 6. LUDS and LSDS were applied in the autumn each year and cultivated in prior to sowing. There was no increase in grain yield following the first

Table 4.   Mean cumulative N recoveries (%) from Table 3
           Figures in brackets indicate numbers of experiments.

	Sludge type		Time of application		Overall
Year	LUDS	LSDS	Spring	Autumn	mean
1	11.9(8)	12.7(8)	12.5(8)	12.1(8)	12.3(16)
2	20.7(8)	17.0(8)	19.6(8)	18.1(8)	18.8(16)
3	23.4(6)	18.4(4)	21.2(7)	21.8(3)	21.4(10)
4	28.2(1)	24.5(2)	26.5(2)	24.2(1)	25.7(3)
5	34.2(1)	29.0(1)	31.6(2)		31.6(2)

Table 5.   Comparison of theoretical half life model with the mean
           cumulative N recoveries observed.

Year	Theoretical	Actual	Difference
1	12.3	12.3	
2	18.4	18.8	+0.4
3	21.5	21.4	-0.1

Table 6.   The efficiency of LUDS relative to N fertiliser following
           successive applications to grassland (7).

		Relative efficiency (%)	
Year	Sludge applied	Yield basis	N Recovery basis
1	yes	7.5	28.8
2	yes	76.2	60.5
3	yes	37.1	35.6
mean 1-3		39.8	40.9
4	no	4.9	6.6
mean 1-4		50.1	55.2

Table 7.   The effects of repeated annual applications of LUDS and LSDS on
           the grain yield of winter wheat (16, 17).

	Year 1		Year 2		Year 3	
	Yield* increase	Yield+ relative to N	Yield* increase	Yield+ relative to N	Yield* increase	Yield+ relative to N
LUDS	0	0	1.83	4.5	1.78	5.2
LSDS	1.85	3.7	2.02	4.3	2.35	5.1
N Fertiliser	3.38	27.0	3.39	27.1	3.39	27.1

* = yield of grain, increase over untreated control (t/ha)
+ = kg grain increase/kg total N applied.

application of LUDS when the response to LSDS was 3.7 kg grain/kg N applied. However, following the second and third applications of LUDS the response increased to 4.5 and 5.2 kg grain/kg N applied respectively when the corresponding values for LSDS were 4.3 and 5.1. The differences in crop performance between LUDS and LSDS treatments are attributable to the higher NH_4-N content of LSDS providing readily available nitrogen to the crop early in the growing season each year when mineralisation of LUDS nitrogen would be too late in the first year for the crop to benefit but was carried over for the following crop.

4. CONCLUSIONS

1. Applying sludge to grassland at different times over the autumn/spring period has relatively small effects on the early growth attributable to winter losses of nitrogen compared with N fertiliser. These effects are not observed in mid and late season growth as the N is then derived from sludge organic matter.

2. Repeated applications of sludge increase the pool of mineralisable N in the soil. This produces progressive increases in mid and late season grass yields relative to early growth.

3. Long term residual effects can be expected from a single application of dewatered sludge. Nitrogen release from both undigested and digested sludges is predictable with fertiliser replacement values decreasing by about half each year.

4. Regular annual applications of liquid sludge have a cumulative effect on residual N values. Experimental data fits closely a theoretical model which indicates that the largest increase in residual value may be expected by the third year of repeated applications.

5. Cropping, climatic and soil conditions affect crop response to sludge, particularly unstabilised sludges. A poor response one year is generally compensated by a large response the following year.

6. By being able to predict residual nitrogen values, further savings in fertiliser may be possible as the amounts of additional fertiliser required could be further reduced where sludge is applied either on a regular basis to farmland or as large one-off applications for land restoration where the long term slow release of nitrogen is very desirable.

5. REFERENCES

1. CATROUX, G., CHAUSSOD, R., GUPTA, S., DE HAAN, S., HALL, J., SUESS, A. and WILLIAMS, J.H. Nitrogen and phosphorus value of sewage sludges. A state of knowledge and practical recommendations. Commission of European Communities, 1982.

2. HALL J.E. Predicting the nitrogen values of sewage sludge. Presented at 3rd International Symposium on Processing and Use of Sewage Sludge, Brighton, UK. Sept. 1983. In press.

3. HALL, J.E. The agricultural value of sewage sludge - a farmers' guide. Water Research Centre, 1984.

4. HALL, J.E. and WILLIAMS, J.H. The use of sewage sludge on arable and grassland. Presented to a joint meeting of WP4 and 5 of the CEC Concerted Action on the Treatment and Use of Sewage Sludge, Uppsala, Sweden, 1983, D Reidel, 1984.

5. HALL, J.E., CARLTON-SMITH, C.H., DAVIS, R.D. and COKER, E.G. Field investigations into the manurial value of lagoon-matured digested sewage sludge. Water Research Centre report 510-M, 1983.

6. Ibid. Field investigations into the manurial value of liquid undigested sewage sludge. Water Research Centre 652-M, 1983.

7. COKER, E.G., HODGSON, D.R. and SMITH, A.T. The effects of undigested primary sewage sludge on the growth and nitrogen uptake of barley and permanent grass. Water Research Centre report 698-M, 1984.

8. COKER, E.G., DAVIS, R.D., HALL, J.E. and CARLTON-SMITH C.H. Field experiments on the use of consolidated sewage sludge for land reclamation: effects on crop yield and composition and soil conditions, 1976-1981. Water Research Centre Technical Report 183, 1982.

9. EDGAR, K.F., FRAME, J. and HARKESS, R.D. The manurial value of liquid, anaerobically-digested sewage sludge on grassland in the west of Scotland. I. Nitrogen value of slude applied in the winter months. Journal of Agricultural Science, Cambridge. In press.

10. COKER, E.G. The value of liquid digested sewage sludge. Parts I, II and III. Journal of Agricultural Science, Cambridge. 1966, 67, 91-107.

11. EDGAR, K., HARKESS, R.D. and FRAME, J. The manurial value of liquid anaerobically digested sludge on grassland in the west of Scotland. Paper prsented to the XIV International Grassland Congress, Lexington, USA, 1981.

12. DE HAAN, S. Effect of nitrogen in sewage sludge on nitrogen in crops and drainage. Presented to a meeting of WP4 of CEC Concerted Action on the Treatment and Use of Sewage Sludge. Dijon, France, 1979. CEC, 1981.

13. KEENEY, D.R., REE, R.W. and WALSH, L.M. Guidelines for the application of waste-water sludge to agricultural land in Wisconsin. Technical Bulletin 88, Wisconsin Department of Natural Resources, Madison, Wisconsin, USA, 1975.

14. KERR, S.N., SOPPER, W.E. and EDGERTON, B.R. Reclaiming anthracite refuse banks with heat-dried sludge. In Utilisation of Municipal Sewage Effluent and Sludge on Forest and Disturbed Land. Sopper, W.E. and Kerr, S.N. (eds). The Pennsylvania State University Press, USA. 1979.

15. PRATT, P.F., BROADBENT, F.E. and MARTIN L.R. Using organic waste as nitrogen fertilisers. California Agriculture, 1973, 27, 10-13.

16. HALL, J.E. The effect of sewage sludge on the growth and composition of winter wheat. First year results. Water Research Centre report 386-M, 1982.

17. Ibid. Second year results. Water Research Centre report 615-M, 1983.

LONG-TERM EFFECTS OF FARM SLURRIES APPLICATIONS IN THE NETHERLANDS

L.C.N. DE LA LANDE CREMER
Institute for Soil Fertility, Haren (Gr.), The Netherlands

Summary

Animal effluents are very valuable resources to supply crops with
minerals and to maintain or improve soil fertility. The amounts to be
applied however should be restricted depending on factors like fertility status of the soil, maximum yield, crop quality and environmental
requirements. The production of biogas offers another possibility for
utilization but does not bring about any change in the quantity of
minerals to dispose of and therefore offers no solution in surplus
situations. Further, not all kinds of manures are suitable for blend-
ing with animal feeds, and then again this is often not permitted.
Finally, excessive applications are hostile to the environment.
The chance that animal effluents change into a useless pool of waste
increases as more livestock is kept independent of homegrown fodder
production.
This paper presents a short survey of results of current agronomic
research on farm slurries under intensive livestock farming conditions
in The Netherlands.

1. INTRODUCTION

A modern livestock enterprise can be operated independently of a feed
supply from its own soil. Intensification of a stock farm on this basis may
vary from maintaining a livestock density that is too high for the available land area to keeping stock on a farm that has no land at all. Of the
minerals entering the farm as constituents of the feed purchased, 80 to 90%
of the major elements and 90 to 95% or more of the trace elements are
excreted in the animal manure. This waste product derives its value as a
fertilizer from the minerals. When a stock farm has no, or insufficient,
land where the manure can be utilized, an excess of minerals arises if the
surplus manure is not removed. This affects, qualitatively and quantitatively, soil, crop, and environment. Such negative effects can be avoided
only by striking a balance between the supply of minerals in the manure and
the amounts needed for production; in doing so, all other fertilizing
agents, such as mineral fertilizer, sewage sludge, etc., should be taken
into account. Thus, increasing the stock density by means of concentrated
feeds has a competitive effect on the purchase of fertilizers for the farm.
Annual consumption of concentrates in the cattle farming enterprises in The
Netherlands increased from 500 kg per cow (with associated young cattle) in
1958 to more than 2300 kg in 1980. About half the minerals contained in the
latter amount is sufficient to compensate for the minerals removed in milk,
meat, and due to leaching. The other half already constitutes a surplus!
Since 1950, the amount of manure produced annually by the entire Dutch
livestock population has doubled to more than 26 t/ha in 1982 (slurries and
solid poultry manure). In large areas surpluses are produced, which leads
to excessive applications and reduction in quality of crop, soil, ground-
water and environment. For the last 15 years, agricultural research in The

Netherlands has been investigating the borderline between manuring and
dumping and has attempted to solve the problem of manure surpluses. This
research was partly subsidized by the EC.

2. FARM SLURRIES

Animal manures comprise a large number of different products. In this
report, only farm slurries will be discussed. They are made up of animal
urine and feces that are stored together and with various amounts of waste
water. In the slurry system virtually no litter materials are used.

Slurries are liquid organic mixed manures, stored under anaerobic
conditions, and containing a multitude of macro- and micronutrients in
addition to organic material. The nutrient concentration varies widely with
type of animal, feed used, and dilution (Table I). There is a correlation,
with a wide variation, between nutrient concentration and dry-matter

Table I. Variation in composition of some slurries, in % of dry matter (6).

	Cattle slurry	Pig slurry
N	5.00 (4.00- 8.00)	8.25 (5.40-14.40)
P_2O_5	2.40 (1.40- 3.30)	5.80 (4.50- 8.30)
K_2O	6.80 (3.50-11.70)	5.00 (2.50- 7.60)
CaO	2.60 (1.70- 3.50)	4.20 (3.00- 5.50)
MgO	1.15 (0.85- 1.65)	1.40 (0.80- 2.50)
Cl	3.15 (1.90- 4.40)	2.50 (0.80- 4.20)

content of slurry that may vary from 2 to 15%. There is a reasonably good
correlation between specific gravity of slurries and their dry-matter
contents. Thus, determination of the specific gravity is insufficiently
accurate as a means to establish the fertilizing value of a slurry.

In Table II the average composition of farm slurries is compared with
that of sewage sludge for agricultural purposes.

Table II. Chemical composition of farm slurries and sewage sludge for
agricultural purposes (5). Contents (% or mg/kg) on a dry-matter basis.

Type of slurry	Fattening calves	Pigs	Cattle	Poultry	Sewage sludge
% dry matter, fresh	2.0	7.8	9.7	14.2	5.0
% organic matter	75.0	66.9	62.9	67.6	62.5
% N	15.0	7.04	4.63	6.48	5.30
% NH_3-N	4.50	3.84	2.32	3.87	2.65*
% P_2O_5	6.50	6.02	2.10	5.63	5.17
% K_2O	12.00	6.40	5.25	4.37	0.61
% CaO	7.00	6.40	2.10	9.65	5.33
% MgO	6.50	1.92	1.05	1.27	0.76
mg/kg Cd	1.50	0.64	0.72	1.06	4.60
mg/kg Cu	95.45	640.00*	43.26	84.48	497.00
mg/kg Hg	0.05	0.05	< 0.10	0.06	2.10
mg/kg Ni	1.50	15.36	4.33	9.50	39.50
mg/kg Pb	6.50	8.45	11.33	4.37	306.50
mg/kg Zn	1425.00	540.00*	205.18	509.00	1480.00

* calculated for 1984

In general, sewage sludge supplies more heavy metals per unit of dry matter than farm slurries.

The fertilizing value of slurry is determined by the kind and amount of nutrients in the feed. By proper selection of feed constituents and optimum dosing of additives (e.g. P, Cu, Zn), manufacturers of mixed feeds can help to keep the contents of less desirable constituents in slurries as low as possible (1).

3. EFFICIENCY OF NUTRIENTS IN FARM SLURRIES

The efficiency of nutrients in farm slurries is not always the same as that of an equal amount of mineral fertilizer. Equal efficiency applies only to the mineral and mineralizable fraction. Of the organic fraction, 50% is mineralized in the year of application, the remaining 50% in subsequent years (aftereffect).

In slurry, the mineral nitrogen fraction consists mainly of ammonia, varying from 50% of the total amount of nitrogen in cattle slurry to 60, 70 and 90% in pig slurry, poultry slurry and slurry from fattening calves, respectively.

Ammonia nitrogen can, partly or completely, volatilize during storage or application. Nitrogen can also be lost due to runoff as a consequence of the often unfavourable weather in the season of application, and as a consequence of heavy rates of application. The severity of the losses depends on a number of factors, some of which can be controlled:

method of storage	- open/covered
method of treatment	- stirring/aerating
method of application	- surface/injection/irrigation
weather conditions	- frost, cool weather, cloudy, rain, sun, dry, windy
time of application	- autumn, winter, spring, summer
promptness in incorporation	- left on surface, ploughing down, injecting

use of nitrification inhibitors
length of the crop's growing period

Figure 1 demonstrates the relation between the nitrogen efficiency of poultry and pig slurries and their time of application, expressed in the amount of precipitation between the time of application and 1 April following (3).

The length of the growing season is particularly important for the efficiency of the organic nitrogen mineralized in the year of application. The part of the mineral nitrogen oxidized to nitrate that cannot be used by the crop will be partly lost due to leaching as a consequence of winter precipitation (Figure 1). Also denitrification and temporary biological fixation partly determine the efficiency of slurry nitrogen. In terms of efficiency, the nitrogen in slurry is therefore never equivalent to a similar content of nitrogen in mineral fertilizer. However, in surplus areas where organic manure is applied more frequently, the nitrogen efficiency of the slurry will improve in the long term due to cumulation of first-year effects and aftereffects (Table III).

Upon intensive use, the initially large differences in nitrogen efficiency among the slurries disappear in the long term.

The nitrogen efficiency of organic manures is also determined by the amount of mineral nitrogen present in the root zone (60 cm) in spring.

Oxidation of ammonium nitrogen (NH_4-N) to ammonia (volatilization) and nitrate (leaching) can be prevented by means of a nitrification inhibitor,

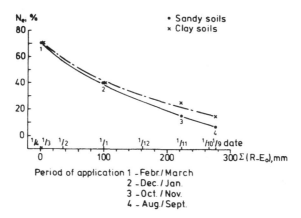

Figure 1. N-efficiency index (N_{ei}) of pig slurry and poultry slurry on sandy soils and clay soils as a function of the surplus precipitation (rainfall R-evapotranspiration E_o) between time of application and 1 April, without winter cover crop (Dutch German data) (3).

Table III. Efficiency and cumulative efficiency of farm slurries following annual application in spring on arable land, in kg fertilizer-N per 100 kg slurry-N per ha (2).

| | Farm slurries from | | | |
	fattening calves	pigs	cattle	poultry
1st year of application	70	55	50	65
equilibrium	77	72	71	76

but the soil temperature then has to remain below 5 °C. In this way heavy losses of nitrogen from autumn-applied slurry can be prevented during winter; this nitrogen becomes available to the crop again in the following spring, so its efficiency is improved.

The nitrogen efficiency of liquid sewage sludge is estimated at 40% in the first year (5).

The efficiency of the other nutrients in slurry is usually considered equal to that of fertilizer. However, the efficiency of phosphate is being reconsidered, because part of it is present in the organic fraction also and will become available only after mineralization.

The direct effect of cattle slurry on arable land is considered to be 60% and on grassland 80% (8). Of the other slurries it is not yet known whether their P-efficiency should be modified. Solid pig manure and broiler manure equal the P-efficiency of superphosphate in the long run, poultry manure on grassland only. The P-efficiency of liquid sludge has been set at 60% (5).

Changes in the pH of the soil can be calculated using the CaO-equivalent values of the farm slurries (9).

4. YIELDS

Slurries contribute to an increase in soil fertility and productivity. They are subject to the same natural laws as other fertilizing agents. Like other organic fertilizer materials, slurries in combination with mineral fertilizer give an attractive increase in production which cannot always be obtained with fertilizer alone. This so-called "rest-effect" is a poorly understood phenomenon, because too many different factors are involved.

5. CROP QUALITY

As is the case with fertilizers, the use of animal manures, or a combination of the two can improve, maintain or reduce crop quality; success depends on determining the correct amount of nutrients to be applied, taking into account the supply of other fertilizing agents (fertilizer, sewage sludge). Excess is harmful to quality and may cause:
* lodging of cereals and flax (N)
* damage of the sward
* unfavourable leaf/cob ratio in silage maize (N)
* change in botanical composition and quality of the sward
* reduction in sugar content and sap purity in sugar beet (N, K, Na)
* decrease in starch content of potatoes (N, K, Cl)
* too high a nitrate and potassium content and reduced Ca and Mg contents of grass
* too high a pH of the soil (poultry manures)
* Cu-accumulation
* increased nitrate losses from the soil
* increased runoff and leaching of P and organic matter
* reduction of worm activity.

To avoid such effects, the proper rate of application should be chosen, taking into account
* composition and nutrient efficiency
* method and time of application
* crop requirement and amounts of nutrients present in the soil
* fertilizing agents available
* soil physical properties (rate of infiltration)
* soil chemical properties
* topography of the terrain.
The first limiting element will determine the rate of application. Elements then in insufficient supply can be supplemented with simple fertilizers.

6. SOIL ENRICHMENT

When manures are applied in amounts that exceed crop and soil requirement - a situation that is rapidly created when surpluses exist (dumping) - elements that are not utilized will accumulate in the soil. Slightly soluble elements accumulate in the topsoil, more mobile elements are dispersed throughout the profile, enrich the subsoil and subsequently also the shallow and deep ground-water. Also, relatively immobile elements (e.g. P) do not remain stationary indefinitely, but gradually move to deeper layers. The speed of transport depends on the magnitude of the surplus and on the type of soil. In peaty and light soils, transport will be much more rapid than in clay soils.

Accumulation of certain minerals in the first few centimeters of soil may be undesirable in view of the possibility of runoff to surface waters.

On grassland, accumulation of Cu is harmful to sheep and reduces worm activity, the latter also on arable land.

7. EUTROPHICATION OF GROUND-WATER

Enrichment of the soil leads to eutrophication of the ground-water. Figure 2 and Table IV demonstrate this for ground-water at 1 m below the surface in a long-term trial with cattle slurry on silage maize on a sandy soil.

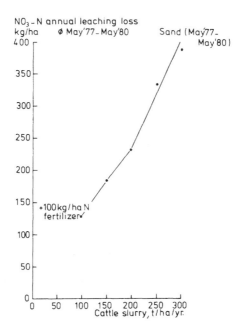

Figure 2. Relation between amount of nitrate (kg/ha NO_3-N) leached to the ground-water and amounts of cattle slurry applied annually since 1973 to silage maize on sandy soil. May 1977 - May 1980 (7).

A leaching percentage in excess of 100 indicates that the soil supply was drawn upon.

8. BIOLOGICAL ASPECTS

Cattle and pig slurries can reduce the development of parasitic nematodes and of root necrosis in silage maize even more effectively than a nematocide (4, 7).

Excessive amounts of pig slurry cause a reduction in the number of worms in grassland. $Cu(HNO_3)$-contents higher than 30-50 mg/kg dry soil can result in diminished reproductive capacity of these animals (10).

90

Table IV. Eutrophication of ground-water as a consequence of different amounts of cattle slurry applied annually since 1973 to silage maize on sandy soil, Nov. 1977 - May 1978. Kg/ha and % of supply in the slurry (7).

Annual application of cattle slurry, t/ha		50	100	150	200	250	300
Leaching loss, kg/ha	Ca^{++}	245	239	321	383	581	679
	Mg^{++}	43	43	64	80	109	113
	Na^{+}	49	83	107	110	208	168
	K^{+}	28	138	174	113	82	153
Leaching loss, % of supply	Ca^{++}	250	148	134	126	132	183
	Mg^{++}	119	62	61	56	50	63
	Na^{+}	120	101	88	67	65	82
	K^{+}	13	31	26	13	7	14

9. LITERATURE

1. Anon., 1975 and 1979. Enkele minerale bestanddelen van mengvoeders in relatie tot de behoefte van de dieren, de uitscheiding in de mest en urine, alsmede enkele gevolgen voor bodem, plant en dier. I. 1975, 72 pp; II. 1979, 70 pp. Werkgroep mineralen in krachtvoer in relatie tot bemesting en milieu. Spelderholt, Beekbergen.
2. Comm. Europ. Commun., Information on Agric. no. 47, 1978. The spreading of animal excrement on utilized agricultural areas of the community, 154 pp.
3. Comm. Europ. Commun. Practical guidelines for the farmer in the EC with respect to utilization of animal manure. In press.
4. Dilz, K., H.W. ten Hag, H.W. Lammers and L.C.N. de la Lande Cremer, 1984. Fertilization of forage maize in the Netherlands. Netherl. Nitrogen Techn. Bull. 14, 25 pp.
5. Haan, S. de, 1983. Landbouwkundige waarde van zuiveringsslib. Sticht. Postacad. Vorming Gezondh. Techn. TH Delft, 44 pp.
6. Lande Cremer, L.C.N. de la, 1978. Bewertung der Wirtschaftsdünger in den Niederlanden, 25 pp. In press.
7. Lande Cremer, L.C.N. de la, et al. 1981. Fragen der Güllerei im Ackerbau und die Umwelt. Bericht 7e Arbeitstagung "Fragen der Güllerei" Gumpenstein, Irdning, Austria, Band I, 201-204.
8. Prummel, J. en H.A. Sissingh, 1983. Fosfaatwerking van dierlijke mest. Bedrijfsontw. 14: 963-966.
9. Sluijsmans, C.M.J., 1970. Einfluss von Düngemittel auf den Kalkzustand des Bodens. Pflanzenernähr. Bodenkunde 126: 97-103.
10. Wei Chung Ma, 1983. Regenwormen als bio-indicators van bodemverontreiniging. Bodembescherming 15, Ministerie VROM, Staatsdrukkerij, 111 pp.

DISCUSSION ON PART I

Chairman (Part I) - Professor Porceddu, Director of CNR Special Research
                       Project, Italy.

Chairman (Part II) - Mr J H Williams, MAFF, United Kingdom.

H TUNNEY to M DEMUYNCK

You state that ammonia loss is high particularly on bare soil. What
evidence do you have that ammonia loss is higher after spreading on bare
soil by comparison with a crop covered soil?

Answer: We have no direct evidence but if there is no crop to absorb the
ammoniacal nitrogen, it is assumed that more will be lost into the
atmosphere from a bare soil surface where temperatures would be higher than
where there is shade from the crop.

B POMMEL to M DEMUYNCK

You stated that 20% of the organic nitrogen from digested sludge was
available, but you did not say over what period of time this result was
obtained?

Answer: Since the organic -N fractions of digested sludges is very
stable, it takes approximately 2 years for 20% of this fraction to become
available to plants.

A SUSS to M DEMUYNCK

What amount of manure do you apply to growing plants?

Answer: The quantities applied correspond to 2-5t/ha of dry solids.

S COPPOLA to M DEMUYNCK

What can you say about the efficiency of thermophilic anaerobic digestion
as regards sanitization and what about the possibility of re-contamination
of such a digested sludge?

Answer: The efficiency of thermophilic anaerobic digestion is complete as
regards sanitization. Indeed the effect on reduction of pathogens is
eventually a function of temperature. Little is known of the possibility
of re-contamination but I would think that the risk is small for a
thermophilic digested waste.

K SMITH to M DEMUYNCK

With reference to the application of treated sludge onto growing crops, was any foliar scorch of crops observed and was any significant difference observed where untreated sludges had been similarly applied?

Answer: To my knowledge there was no crop damage as result of surface spreading of digested sludge. It has been observed that there is a more rapid response in terms of crop growth following the use of digested sludge compared with untreated sludge resulting in an earlier crop harvest.

J VOORBURG to M DEMUYNCK

You found a reduction in pathogens by anaerobic digestion. Can you give the percentage of this reduction, and did you take into account the effect of just the retention time? A period of anaerobic storage of one month usually gives a reduction of pathogen of approximately 90%.

Answer: Reduction is a function of pathogen type. For plant pathogens a reduction of 100% can be obtained and also for parasitic cysts. In the case of helminthic ones, however, the reduction is of the order of only 20-30%. The sanitization effect is a function of retention time and temperature. It would therefore be true that a further period of storage would reduce pathogen numbers still further even though temperatures would be much lower than inside the digester.

J HALL to U TOMATI

I was interested in your comments on the production of plant hormones by soil micro-organisms following the application of sludge. At Rothamsted in the UK, worm worked sludge compost has produced substantially higher plant yields than unworked compost. This effect is much larger than expected from nutrient effects - is it possible that the increase in microbial activity has increased phytohormones in the soil and benefited crop growth?

Answer: When I began to study vermicomposting of organic wastes (sludge, urban wastes...), the action of earthworms and their castings on the plants, my attention was drawn by the rapid root development with plants grown in the presence of earthworms or their castings. There is evidence in the literature, and several experiments at Rothamsted confirm it, that earthworms and their castings greatly enhanced root development and plant growth.

I agree with you that these effects cannot be related entirely to nutrients. Because growth regulators are known to enhance root development, plant growth and yield, I think that microbial growth regulators may play an important role in plant development. My researches on phytohormones and their presence in castings and the action of these on plant growth and root development, support this hypothesis.

We must remember that castings are the result of microbial activities (mineralization and biosynthesis of many active metabolites) in the earthworm gut. Since these activities occur altogether, castings are probably rich both in available nutrients and growth regulators.

K SMITH to U TOMATI

You presented some data giving the nitrate content of plants as affected by sewage treatment. What were the units of concentration used? At what stage of growth was this factor determined?

Answer: Nitrate content is expressed as u moles of nitrate per gramme of fresh weight. Nitrate reductase activity was determined over the whole vegetative cycle of the plant but only the values at plant emergence are reported in the paper.

S DE HAAN (Comment)

The main reason for the differences in nitrogen availability between cattle and pig slurries and between different treatments of the same slurry is, in my opinion, the difference in the rate of decomposition of the organic matter in these products. The first need is a practical method to determine this rate of decomposition of the organic matter in organic soil amendments. At the same time this should also give us a measure of the humus forming capacity of these products.

J VOORBURG to J BESSON

Were the losses of ammonia during aeration caused by volatilization or by nitrification followed by denitrification. This is important because NH_3 volatilization can contribute to the effects of acid precipitation.

Answer: Losses of nitrogen occurred in the form of ammonia volatilization which resulted in a loss of total nitrogen also. This was the only possibility for the nitrogen to be lost from slurry in the treatment process. If it had nitrified appreciably one should find some nitrate -N during or at the end of the treatment process. In fact, that was never the case; we have never measured amounts greater than 50 mg/litre of nitrate -N which is a negligible amount from the agronomic point of view. I cannot say that this would be so in the nitrification/denitrification cycle but during the treatment process, quantitatively this would be minimal in agronomic terms. For myself, the connection between ammonia losses by volatilization and "acid rain" theory is not evident and I do not follw the reasoning behind this.

H TUNNEY to J BESSON

You show a loss of about 9% ammonia from cattle slurry and half that from pig slurry during aeration. However, after landspreading, will you not loose a much higher fraction of the ammonia?

Answer: Losses of nitrogen by volatilisation of ammonia were measured in model experiments. The hypothesis was that, if aerated slurries contain the least NH_4-N, slurries treated by anaerobic digestion the most NH_4-N and stored slurries an intermediate amount, one must expect that the nitrogen losses by volatilisation of ammonia during landspreading will be in the order of aerated slurries < stored slurries < anaerobically digested slurries. The results confirm this hypothesis but still requires more detailed verification before publication.

J WILLIAMS to P SPALLACCI

Which methods were used in assessing available P and could the interpretation be used with confidence to reduce fertilizer crop requirements?

Answer: The method used was that of Olsen and the data can be used to establish fertilizer needs with reasonable confidence.

S GUPTA to P PAGLIAI

Did you also estimate the distribution of pores in soils according to their size, looking at differences between macro and micro pores?

Answer: Pore sizes were differentiated into 3 classes, viz. less than 50 um (the storage pores), from 50-500 um (the transmission pores) and those larger than 500 um. The increase in porosity of the slurry treated plots was largely due to an increase in pores of 50-500 um in diameter indicating a clear improvement in soil conditions for root growth and water movement.

J HALL to P PAGLIAI

Did you take soil samples for pore measurements at different times of the year? The samples taken at the end of June are only one month after a slurry application - are you sure that the soil had reached equilibrium in that short time?

Answer: Samples were also taken in September and the total porosity showed a decrease compared to June in both treated and untreated plots. This was ascribed to the natural compaction of soils which occurs by crop harvest time. The greatest increase in porosity of slurry treated plots occurred in the first few weeks after slurry application confirming that decomposition of the organic matter was fairly rapid reaching a peak in June when soil moisture and temperature conditions are optimal for biological activity.

A SUSS to P PAGLIAI

Can you offer any explanation for the significant differences in soil structure between the two times of slurry application. Is it the large amount of water applied, is it perhaps a salt effect because the amount of organic matter is always the same?

Answer: This can usually be ascribed to soil biological activity. In June, this was high due to optimal conditions of soil moisture and temperature. Samples from plots treated in June showed the presence of biopores, originated by soil fauna and by carbon dioxide liberated by the decomposition of organic matter. Bioactivity was higher than in plots treated in October or February. It is well known that high biological activity is imporant for the improvement of soil structure.

S de HAAN to P PAGLIAI

In your paper you mention that the decomposition rate of soil organic matter is rather high in Italy. Have you figures to demonstrate this and have you comparable figures from more northern parts of Europe?

Answer: The decomposition rate of soil organic matter is high in Italy firstly because of the climate and secondly because of the intensive cultivations. It is well known that intensive cultivations cause a strong decrease of soil organic matter especially if the organic matter returned from crop residues is less than that returned to the soil from a full vegetation cover. Moreover in Italy there is a severe shortage of farmyard manure due to the large reductions in livestock numbers. I do not have comparable figures from northern parts of Europe but I can say that in the arable lands of Northern Italy the organic matter content is higher than in arable lands of the south where the organic matter in soil is very often less than 1%.

A SUSS to S DE HAAN

Can you explain the negative potassium effect in sugar beet?

The potassium content in sewage sludge is rather low - maybe it is an indirect effect of sewage sludge?

Answer: The mean increase in K in the sugar juice was about 4 meq/100g sugar or 5 kg/ha, whereas in this amount of sludge, about 30 kg K was supplied over and above that given in the mineral fertilizer. The increase in K in the sugar juice, which lowers the extractability of the sugar, can be explained by the extra potassium supplied by the sludge.

S GUPTA to S DE HAAN

Have you considered the effect of boron on the sugar content of beet?

Answer: Averaged over the six soils and the sludge application rates, the amount of water soluble B was increased by 16% at the end of the 6 year experimental period. This rather small increase in the water soluble B content of the soil could not be regarded as a positive effect of sewage sludge.

J WILLIAMS (Comment)

In relation to the detrimental effects of sewage sludge on sugar content and yield there have been instances in UK where historic applications of sewage and sewage sludge had resulted in large reductions in sugar content - from about 18% down to 13%. We ascribe this to uptake of mineral N late in the season from depth in the profile of very sandy soils.

J WILLIAMS to J HALL

It is rather surprising to find that residual effects in 2nd and 3rd year were higher for digested than for untreated sludge.

Answer: The effects shown were all relative and were measured in the presence of fresh applications each year - the apparently better relative effects from digested sludge in the 2nd and 3rd years are due to the higher ammonia content increasing the first harvest in each year.

B POMMEL to J HALL

In the model which you presented to predict percent of cumulative N recovery depending only on the number of years - how would you expect it to vary with other parameters?

Answer: Climatic variations from year to year would probably have the greatest effect but, as illustrated in table 3, time of application, soil type and also cropping have significant effects.

K SMITH to J HALL

Would you be optimistic about the possibility of applying the predictive model for nitrogen recovery in practical farming situations, or do you think that the model needs more refinement before it can be used with any degree of success?

Answer: Longer term trials are really required, particularly to identify the conditions which lead to lower residual effects such as dry conditions or arable cropping, before such a model can be used with confidence. However, as it stands, it does indicate that residual effects can be significant in some instances and are worth taking into consideration by the farmer.

O FURRER to J HALL

Do you know of a simple chemical method to characterize organic matter in order to predict the rate of mineralization of organically bound nitrogen?

Answer: There are a number of rapid chemical incubation tests that have been tried. Whilst they may be useful in relative terms on a laboratory scale they have not been very successful when compared with results from field trials.

A. CUMULATIVE AND RESIDUALS EFFECTS OF SLUDGES AND

FARM SLURRIES

PART II

Long-term field experiments on the fertilizer
value and soils ameliorating properties of
dewatered sludges

The fertilizing value of slurry applied to arable
crops and its residual effects in a long-term
experiment

Experiments on the fertiliser value of animal
waste slurries

Phosphate balance in long-term sewage sludge and
pig slurry fertilized field experiment

Chemical characterization of soil organic matter
in a field study with sewage sludges and composts

Discussion on Part II

LONG-TERM FIELD EXPERIMENTS ON THE FERTILIZER VALUE AND SOILS
AMELIORATING PROPERTIES OF DEWATERED SLUDGES

I. KOSKELA
Agricultural Research Centre, 31600 Jokioinen, Finland

Summary

Field experiments were carried out in the years 1973-1982 at
Agricultural Research Centre, Finland. Highest amount of digested
sewage sludge was 100 ton dry matter per hectare. In addition to
sewage sludge 50 or 100 kg N per hectare in NPK-fertilizer has been
given every spring. The purpose was to use half of the normal
nitrogen amount as mineral fertilizer to find out the effect of the
sludge. It has been analysed both plant and soil materials.

Sewage sludge increased most of all barley yields grown in clay
soil rich in organic matter, still in 1982 700 kg per hectare. Sludge
increased barley yields only few hundred kg per hectare grown in
sandy soil, and only during 3-4 years.

Nitrogen concentrations in the grain were highly dependent upon
sludge treatments. Soil organic matter increased in clay soil with
the sludge treatment.

Concerning the quidelines it should be lower amount of sludge
for sandy soils poor in organic and clay material and higher for
clay and humus soils, because they can absorb elements. So there is
not so much harm for groundwater and also for surface water.

1. INTRODUCTION

Problems connected with sludge disposal, improving sludge handling
techniques and tendencies toward recycling of materials were the main
reasons to start sewage sludge experiments in Agricultural Research
Centre. There was not enough knowledge about sludge fertilizing value in
comparison with manure.

The most important thing for a farmer is to find out, how to have
the best yields and also the best economical value of his own work. The
effects of sewage sludge on crops vary both with soil, crop and year,
that means weather conditions.

Most of sludges in Finland are dewatered by reason of the long
wintertime. Otherwise there are not enough depositary places. Almost half
of the solids of sewage sludge consists of organic compounds. So it was
expected that sewage sludge is an important source of organic matter for
the soil humus supply. On the otherhand sewage sludge is rich in
phosphorus, nitrogen and some metals, both useful and harmful.

2. EXPERIMENTAL

To have more knowledge about agricultural value of sludge field
experiments were started in Agricultural Research Centre near Helsinki
and in seven experimental stations (Figure 1).

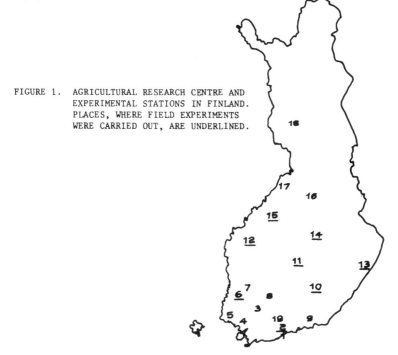

FIGURE 1.   AGRICULTURAL RESEARCH CENTRE AND
            EXPERIMENTAL STATIONS IN FINLAND.
            PLACES, WHERE FIELD EXPERIMENTS
            WERE CARRIED OUT, ARE UNDERLINED.

Four types of sludge resulting from the addition of ferroc sulphate
and lime to remove phosphorus from wastewaters were applied in dewatered
or liquid forms to five different kind of soils in 17 field experiments.
Cumulative element loadings to soils in the experiments varied with sludge
type and application rate but did not exceed 3000 kg N, 2600 kg P and 50
ton organic matter per hectare. Mean mineral element contents of sewage
sludge and manure are presented in Table I.
    Soils were
        - clay rich in organic matter
        - fine sand rich in organic matter
        - fine sand poor in organic matter
        - silt
        - peat.
Soil pH varied between 5,0-6,5.
    Barley, oats, rye, timothy, turnip rape, sugar beet, potato and
carrot were grown during the experimental years 1973-1982. Some
experiments are still going on. The duration of experiments varied between
3-9 years.
    Plant and soil samples were analysed (N, P, K, Ca, Mg, Zn, Cu, Mn,
Pb, Cd and soil humus content in some cases).
    Sludge was applied only in the beginning of experiment. In addition
to sewage sludge 50 or 100 kg N per hectare in NPK-fertilizer has been
given in drills every spring. The purpose was to use half of the normal

TABLE I. MEAN ELEMENT CONTENTS OF SEWAGE SLUDGE AND MANURE IN FRESH MATERIAL.

	Dry matter	Nitrogen total	NH_3-N	P	K	Ca	Mg	Zn	Cu	Mn
						kg/ton				
Sewage sludge										
– dewatered	200	5,5	1	5	0,6	2	1	0,140	0,040	0,090
– liquid	55	1,8	0,7	1,4	0,2	0,6	0,3	0,040	0,010	0,030
Cattle manure										
– with bedding	184	4,8	1,3	1,7	4,2	2,4	0,9	0,040	0,006	0,050
– slurry	81	3,6	2,1	1,0	3,0	1,3	0,5	0,020	0,004	0,020
– urine	26	3,6	3,3	0,2	6,0	0,2	0,2	0,004	0,001	0,003
Pig manure										
– with bedding	230	7,4	2,8	3,7	3,9	5,3	1,4	0,100	0,030	0,060
– slurry	92	7,0	5,1	1,9	2,8	2,8	0,7	0,060	0,030	0,020
– urine	18	4,4	4,0	0,4	3,0	0,4	0,1	0,008	0,003	0,003

nitrogen amount to find out the effect of the sludge.

3. RESULTS

Element concentrations in plant material grown in different experiments varied by reason of soil, crop species, year and sludge (TABLE II). The concentrations were usually higher in vegetal parts of plants than in grain, seed or roots.

Crop effects

It was supposed earlier in Finland that crops with long growing period shall have higher yields with sludge than for example grain. Sewage sludge increased most of all barley yields grown in Tikkurila clay soil, rich in organic matter. In following chapter it will be examining closer field experiments in Tikkurila, durations 9 and 5 years.

In the experiment on sandy soil barley was grown 4 years and one year carrot. On sandy clay soil it has been every year four different blocks (I-IV):

Year	I	II	III	IV
1974	rye	sugar beet	turnip rape	barley
1975	barley	" "	" "	"
1976	hay	" "	" "	"
1977	"	" "	" "	"
1978	"	" "	" "	"
1979	"	" "	" "	"
1980	barley	barley	barley	"
1981	"	"	"	"
1982	"	"	"	"

First year in 1974 there was too much nitrogen in clay soil with sludge treatments and the straw was not strong enough (Figure 2). Following years sludge application increased very significantly barley yields, still in 1982 700 kg per hectare. Barley grain contained especially more nitrogen and zinc with sludge treatments (Table III).

TABLE III. UPTAKE OF NUTRIENTS (KG/HA) IN THE YEARS 1974-1982 BY BARLEY GRAIN YIELDS, STRAW WAS COLLECTED AWAY BUT NOT WEIGHED.

TON/HA D.M.		KG/HA							
SLUDGE	YIELDS	N	P	K	Ca	Mg	Zn	Cu	Mn
0	29 173	409	88	138	10	29	0,90	0,18	0,44
50	36 762	623	113	172	13	37	1,81	0,24	0,61
100	38 567	706	117	179	14	38	2,13	0,27	0,68

TABLE II. LOWEST AND HIGHEST GRAIN ELEMENT CONCENTRATIONS RELATED TO DIFFERENCES BETWEEN SOILS AND CROP SPECIES, OTHER TREATMENTS WERE SAME.

Soil	Crop	Treatment	g/kg dry matter				
			N	P	K	Ca	Mg
Clay, rich in humus	barley	no sludge	13,40-24,10	2,00-5,00	3,60-7,00	0,30-0,50	1,10-1,40
		sludge	15,40-25,90	1,60-5,40	3,50-7,40	0,20-0,60	1,20-1,40
Fine sand	barley	no sludge	13,10-28,60	1,80-4,90	3,80-6,50	0,20-0,60	0,70-1,30
		sludge	12,80-30,20	1,80-4,80	3,40-6,80	0,30-0,70	0,70-1,30
Silt	barley	no sludge	14,30-22,70	2,80-4,80	4,60-6,50	0,30-0,60	1,00-1,50
		sludge	15,20-28,20	3,10-5,10	4,50-7,10	0,30-0,60	1,00-1,50
Fine sand, rich in humus	barley	no sludge	11,90-19,60	2,30-3,70	3,50-6,50	0,30-0,50	0,80-1,20
		sludge	14,80-32,30	2,10-5,00	3,50-7,60	0,30-0,50	0,80-1,30
Clay, rich in humus	oats	no sludge	17,10-28,20	2,40-4,30	2,80-7,20	0,40-0,80	0,90-1,60
		sludge	17,60-29,20	2,00-4,30	2,90-7,70	0,40-0,70	0,90-1,50
Fine sand	oats	no sludge	19,20-27,50	3,50-4,60	3,90-8,10	0,50-0,80	1,30-1,60
		sludge	19,30-28,70	3,40-4,70	3,80-7,90	0,50-0,90	1,10-1,40
Silt	oats	no sludge	17,70-24,60	2,20-4,50	3,00-5,00	0,40-0,70	0,80-1,40
		sludge	16,20-24,30	2,20-4,90	2,70-5,00	0,40-0,70	0,80-1,40
Clay, rich in humus	rye	no sludge	18,20-20,50	2,50-2,90	4,50-4,80	0,40-0,40	1,00-1,10
		sludge	22,20-24,10	2,80-3,20	4,60-5,20	0,40-0,50	1,10-1,20

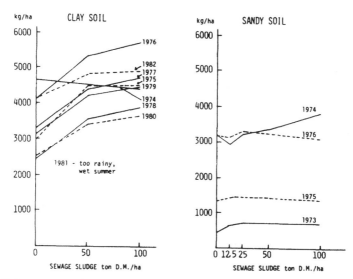

FIGURE 2.   Barley yields in sandy soil and in clay soil, kg/ha. Sludge
was given only in the beginning of the experiments. Yearly
50 kg N/ha was given as NPK-fertilizer.

Sewage sludge increased barley yields grown in sandy soil only few
hundred kg per hectare and only during 3-4 years. Differences between
soils were significant.

The yield level in sandy soil was much lower and depended more on
weather conditions, because the water holding and cation exchange
capasities in that soil were very low. In the years 1974 and 1976 there
was enough rainy days in Tikkurila.

It was not so easy with other plants. For sugar beet the amount of
sludge was so big, that the seeds went too deep and were grown much slower.
Amount of nitrogen was also too high and therefore in autumn sugar
concentration was low.

It was difficult to balance mineral fertilizing for many years for
sugar beet and hay, because they need plenty of potassium. Otherwise it
was aim to have same treatments for all four blocks. In Figure 3 there
are shown mean values of yearly uptake of nutrients by different plants.
Unfortunately straw yields were not weighed. All plant material was
collected away from the fields every year.

It has shown in figure 3 that hay and sugar beet uptake more
potassium than cereals and turnip rape.

Yearly variation

Yields variations and yearly uptake of nutrients varied quite a lot
during the experimental years (Figure 4). In the years 1973-1982 there
were warm and cold summertimes, dry and rainy harvest seasons. Sometimes
it was difficulties to harvest just in right time. If the springtime was
dry, it was possible to see dark colour in the soil there, where sludge
was supplied in clay soil. Its water holding capacity was good.

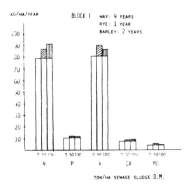

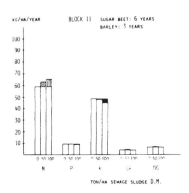

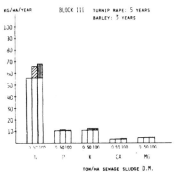

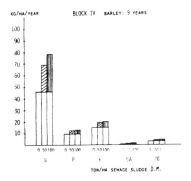

FIGURE 3. Yearly uptake of nutrients in different blocks, mean values of grain, beets, hay and seeds.

Soil effects

The soils in field experiments were chosen to represent different Finnish textures: fine sand, silt and clay.

Element concentrations in the plant materials were related to soil texture and to soil organic matter. If the soil was poor of organic matter and for example fine sand, most often the element concentrations were high in plant material even without sludge application and increased with sludge treatment. The effect of sludge application was very short.

In the soils with high clay and humus contents sludge phosphorus is clearly seen even after 9 years (Figure 5). Shortage of potassium is clear, too. It should take care about potassium fertilizing, if the soil is poor, when sewage sludge is used.

Soil organic matter

In the year 1977, when the sandy soil experiment was over, there was no possibility to analyse soil organic matter. In the years 1980-1982 soil organic matter in sandy clay soil was analysed (Figure 6). Differences between crops and treatments are significant.

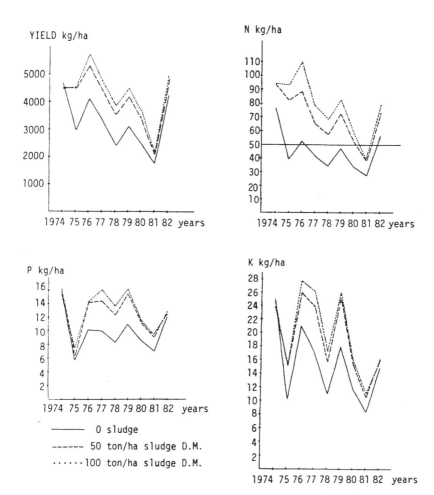

FIGURE 4.   Yield variations and yearly uptake of nutrients in barley
blocks during the years 1974–1982.

4. DISCUSSION AND CONCLUSIONS

Dewatered sewage sludge is very valuable matter for soil. It could
have very positivy effect on yield amounts and also on the concentrations
of elements. The effect varies quite much depending on soil, crop and
year. Most difficult is to find out economically best way to use sludge.

The quidelines prefer small sludge applications per year. Anyway, it
seems that 40-50 tons dry matter per hectare should give the best
economical result when used dewatered sludges in Finland for good soils.

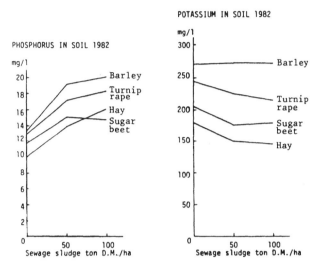

FIGURE 5. Phosphorus and potassium concentrations in the soil 9 years after sludge application.

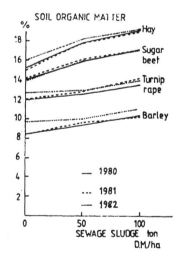

FIGURE 6. Soil humus contents 7-9 years after sludge application. In clay soil.

It was surprising that differences between soils were so high. Concerning the quidelines, it seems that it is no reason to use same amounts sludge for all kind of soils. It should be lower amounts sewage sludge for mineral soils poor of organic and clay material and higher for clay and humus soils, because they can absorb nutrients. So we can be sure we don't have so much difficulties for the pollution of groundwater and surface water.

The most important thing is to eliminate all harmful elements before purification plants.

REFERENCES

1. COKER, E.G., DAVIS, R.D., HALL, J.E. and CARLTON-SMITH, C.H. 1982. Field experiments on the use of consolidated sewage sludge for land reclamation: effects on crop yield and composition and soil conditions, 1976-1981. Technical Report TR 183, Water Research Centre.
2. GUIDI, G. and HALL, J.E. 1983. Effects of sewage sludge on the physical and chemical properties of soils. Third Intern. Symp. on the Processing and Use of Sewage Sludge. Brighton, United Kingdom.
3. Puhdistamolietteen käyttöopas. 1984. Tieto tuottamaan 33, MKL. Finland

THE FERTILIZING VALUE OF SLURRY APPLIED
TO ARABLE CROPS AND ITS RESIDUAL EFFECTS
IN A LONG-TERM EXPERIMENT

J.P. DESTAIN, Y. RAIMOND and M. DARCHEVILLE
Station de Chimie et de Physique agricoles
Centre de Recherches Agronomiques - Gembloux (Belgique)

Summary
The agronomic evaluation of livestock effluents has been
studied at the Agricultural Research Centre of Gembloux
as part of a large-scale joint programme (EEC contract
N° 251). Therefore numerous experimental fields have
been set up in the main agricultural regions of Belgium.
One of these experiments has been carried out in 1974 in
Gembloux, in a loamy region, with the aim of measuring
the long-term efficiency of slurry N applied to arable
crops on the one hand, to look at the slurry as the only
potential source of N ensuring a good decomposition of
crop residues turned under on the other hand. By taking
advantage of an existing experimental lay-out, it has
been possible to evaluate the overall fertilizing value
of slurry considering its action on the nutrient content
and on some physical characteristics of the soil.

1. Introduction
     The agronomic evaluation of livestock effluents has been
the subject of a large-scale joint research programme at the
Agricultural Research Centre of Gembloux (EEC contract N° 251).
The results have been gathered mainly from annual experiments
conducted in the different agricultural regions of Belgium (1).
An experiment has been carried out in 1974 in Gembloux, in a
loamy region with the aim of measuring the long-term efficiency
of slurry N applied to field crops (sugar beet and cereals) on
the one hand, to study the possibilities of slurry applications
as the only source of N ensuring a good decomposition of crop
residues turned under on the other hand. By taking advantage
of an existing experimental lay-out, it seemed interesting to
estimate the overall fertilizing value of slurries with regard
to their action on the nutritional reserves and on some physical properties of the soil.

2. Experimental lay-out
     The experimental field was established at Gembloux on a
deep loamy soil, slightly gleyified and locally waterlogged
during winter.

The three-course rotation investigated at Gemboux included sugar beet followed by two small grains namely winter
wheat and barley with oats the first year of the experiment.
Greens and straw have been turned under and vetch ploughed
in as a green manure. The four experimental treatments investigated have been replicated 6 times following the ran-
domized block lay-out and are described hereunder.
Treatment I : Nitrogen fertilizer applied as urea or ammonium
nitrate. The total N dressing is calculated before sowing
and gives a good picture of the practical average level.
This fertilizer dressing is split when applied to small
grains.
Treatment II : The N required for the decomposition of straw
is applied as slurry N; the rest as fertilizer N and as a
split dressing for cereals. The total N applied in treatment
II is the same as in treatment I.
Treatment III : The whole of the N dressing is applied as
slurry N; the N dressing is based on the chemical analysis
of the slurry and is equivalent to the amount N as applied
in treatments I and II.
Treatment IV : Nitrogen is applied wholly as slurry N; the
total N dressing amounts approximately to 150 % of the N
applications in the three above-mentioned treatments.
The experimental lay-out of the 1975-1983 field experiments
is shown in table I; the chemical composition and the origin
of the slurry are given in table II.

2. Crop yields and efficiency of slurry N
2.1. Preliminary comment

    Nitrogen is the most important nutrient conditioning
mostly the maximum possible yield; the least excess or
deficiency are made very soon visible.
    P and K reserves are particularly large in most of the
Belgian soils. Consequently, crop yields are practically
linked to the availability of soil and fertilizer N (2).
The use of slurry as the only source of nitrogen for crops
maked in necessary to reduce losses to a maximum during and
after spreading.
More than half of slurry N is NH_4^+-N (3) and liable to
volatilize if not immediately ploughed in after spreading.
The rest is organic N and will undergo nitrification, as
well as ammonium, N after having been mineralized more or
less quickly according to the climatic conditions, and made
available to the plant.
Leaching of nitrate N by winter rains (4) (5) and losses
through denitrification (6) compote with its uptake by the
plant.
2.2. Crop yields

    Table III indicates these yields and their statistical
signifiance. Figure I shows the evolution of the relative
yield values as compared to treatment I.

TABLE I - Gembloux Experimental Field - Fertilizer and
slurry N dressings (kg ha^{-1}).

Year	Crops	Treatment I		Treatment II		Treatment III		Treatment IV	
1975	Oa	S Cr T	20 92.5 112.5	S C T	20C 92.5 112.5	S C T	20C 75C+37.5 132.5C	S C T	27C 100C+37.5 164.5C
1976	W B	S Cr T	40 60 100	S C T	40C 60 100	 T	 100C	 T	 160C
1977	S B	S Cr T	34 100 134	S C T	34C 100 134	S C T	34C 100C 134C	S C T	53C 150C 203C
1978	W W	S Cr T	0 120 120	S C T	0 120 120	 T	 122C	 T	 174C
1979	W B	S Cr T	40 80 120	S C T	52C 80 132	 T	 137C	 T	 172C
1980	S B	S Cr T	46 153 199	S C T	33.5C+20 153 206.5	S C T	53C 144C 197C	S C T	77C 205C 282C
1981	W W	S Cr T	0 119 119	S C T	0 119 119	 T	 126C	 T	 185C
1982	W B	S Cr T	60 105 165	S C T	31P+30 105 166	 T	 128P	 T	 203P
1983	S B	S Cr T	40 136 176	S C T	47P 136 183	S C T	47P 221P 268P	S C T	57P 356P 413P

Oa   : Oats
W B : Winter Barley
W W : Winter Wheat
S B : Sugar beet

C : N from cattle slurry
P : N from pig slurry

S : N required for straw
     decomposition
Cr : N required for crop
T : Total N dressing

Remark - Fertilizer N is applied as split dressings to cereals during the
growing period.

TABLE II - Chemical composition of spread slurries
(wet matter basis).

	Origin	Years of spreading	N $°/_{oo}$	P_2O_5 $°/_{oo}$	K_2O $°/_{oo}$
Slurry 1	C	1975	4.03	1.31	3.50
Slurry 2	C	1976 - 1977 1978 - 1979 1981	4.00	1.00	5.90
Slurry 3	C	1980	2.91	0.70	2.88
Slurry 4	P	1982 - 1983	5.25	4.00	4.00

TABLE III - Crop yields (kg grain 16 % moist. ha^{-1} - kg sugar ha^{-1}).

Year	Crops	Treatment I	Treatment II	Treatment III	Treatment IV
1975	Oa	$3,378^b$ *	$3,117^b$	$2,893^a$	$3,343^b$
1976	W B	$7,242^b$	$7,182^b$	$6,443^a$	$6,909^b$
1977	S B	$7,999^a$	$7,890^a$	$7,224^a$	$8,044^a$
1978	W W	$5,839^a$	$5,659^a$	$5,528^a$	$5,725^a$
1979	W B	$6,158^d$	$5,492^c$	$4,172^a$	$4,626^b$
1980	S B	$8,710^a$	$7,846^a$	$7,392^a$	$7,930^a$
1981	W W	$7,645^b$	$7,572^b$	$6,211^a$	$6,620^a$
1982	W B	$7,730^c$	$7,289^b$	$4,986^a$	$5,250^a$
1983	S B	$10,879^a$	$9,916^a$	$10,282^a$	$9,788^a$

* Between-treatment values followed by the same letter
are not significantly different at the 0.05 probability
level.

2.2.1. Sugar beet

The spreading of the slurry has been carried out just before sowing. Yields of sugar in kg ha^{-1} are given for 1977, 80 and 83. No statistically significant differences could be found because of the large values of the variation coefficient.

The results can be summed up as follows :
In 1977, the application of 134 kg N ha^{-1} as slurry N cannot meet the needs of the crop in treatment III which shows the lowest field. The maximum yield observed in an appended experiment results from an application of 130 kg ha^{-1} as mineral fertilizer N.
In 1980, the yield of treatment III (197 kg slurry N ha^{-1}) is smaller than that of treatment IV (282 kg mineral fertilizer N ha^{-1}) notwithstanding its N dressing exceeding by far the maximum 140 kg.
Finally in 1983, treatment III with an application of 268 kg ha^{-1} as slurry N gives a very satisfactory yield higher than that of treatment IV where an excess fertilizer N (413 kg ha^{-1}) has lead to a non significant decrease in the sugar content of the roots.
It seems that the efficiency of slurry N is rather unsteady and not always so high as expected, even after Spring spreadings. The value of slurry as far as the decomposition of straw in concerned is difficult to calculate as this microbial action starts in early Autumn in plots with green manuring. It seems however that slurry N is inferior to fertilizer N if treatments II and I are compared.

2.2.2. Winter wheat

Results are given for the 1978 and 1981 crops.
In 1978, the maximum possible yield was only 5,800 kg ha^{-1} with a fertilizer dressing of 80 kg N ha^{-1}. This low yield results from bad weather conditions : late sowing, cool Summer. The slurry treatments were inferior to treatment I but the differences were not statistically significant.
1981 was a crop year of high production : over 7,500 kg ha^{-1} corresponding to a 120 kg ha^{-1} mineral N dressing. Yields of treatments III and IV were significantly lower than the values found for treatments I and II notwithstanding equivalent (126 kg ha^{-1} treatment III) or larger N applications (185 kg ha^{-1} treatment IV).

It can be assumed that the slurry N applied before sowing has been partially nitrified in Autumn and that some of the nitrate N has been leached down by heavy Winter rainfall (7). Nitrogen is badly needed by the plant at growth resumption and especially at stem elongation; however the rooting system is not yet suffisantly developed at this stage of plant development to reach the nitrogen wave having leached down the soil profile. The nitrogen stress is thus very acute (plant yellowing) and crop yields made hazardous.

Generally, a winter wheat crop does not require any
nitrogen application up to tillering excepting in the case
of an early sowing (1981) owing to its limited root development before the onset of winter on the one hand, considering
the nitrogen released by the sugar beet greens turned under
on the other hand.

2.2.3. Winter barley

Winter barley was grown in 1976, 1979 and 1982.
Slurry N (treatment III and IV) leads generally to yields
which are very significantly lower than those of fertilizer N
(treatment I). The extra slurry application in treatment IV
makes up only partially for the relative inefficiency of
slurry N.
The plant yellowing resulting from the lack of available N
at the beginning of stem elongation as well as the reduction
in grain yield are made still more conspicuous than in the
case of a winter wheat crop. On the other hand, the winter
barley sown in early autumn has developed numerous tillers
before winter sets in. Therefore it is essential to apply
a nitrogen dressing as a starter for the breakdown of straw
residues turned under before sowing. Slurry N seems not
to be so efficient as mineral fertilizer N for this purpose
(compare yields in treatments II and I).

2.2.4. Oats

Oats was sown only at the beginning of the experiment
in 1975 after the barley crop. Yields were very low and the
lowest for treatment III.

2.3. Efficiency of slurry N

The efficiency of slurry N is the ratio between its
nitrogen effect and the total amount N in the applied
dressing, the nitrogen effect being measured by comparing
the yields of slurry N to the yield curve of a reference
nitrogen fertilizer (see appended test).
Table IV shows the nitrogen efficiency values of the slurry
in treatments III and IV as calculated for several crops.
Missing values are related to yield data diverging too much
from the reference curve in the appended test. The efficiency of slurry N as calculated for winter cereals is very
low, often less than 40 %. On the other hand, the efficiency
values calculated for a sugar beet crop varies between 35
and 66 %, the higher values being related to spring spreadings. Unfortunatelly wet soil conditions impede very often
spring applications. The low recovery of slurry N by winter
wheat and barley results from its quick nitrification in
autumn when plants are unable to take up the formed nitrate N,
a fraction of which being leached down by winter rains.
The wave of nitrate N in the soil profile is too deep to be
absorbed by plant roots at the resumption of growth which
leads to a nitrogen shortage. There is a direct relationship
between low yield and recovery values of slurry N on the
one hand, high winter rainfall on the other hand.

This is illustrated by the rainy winter periods of the 79, 81 and 82 cropping seasons - more than 350 mm recorded between the beginning of November and the end of March - and the less than 250 mm value recorded in 1976. It should be noted that an excess slurry application as in treatment IV does not improve the efficiency values. On the contrary, this slurry N may depress the yield (sugar beets - 1983) as already mentioned by Lecomte et al. (1) in their annual experiences. Generally the efficiency values as shown in table IV are lower than those calculated by other authors from formulae based on the time of spreading and on the ratio between mineral N and total N in slurry (8). These discrepancies result probably from differences in the experimental conditions as far as soil water characteristics and crop management practices are concerned : waterlogging of the experimental plots during the winter period increasing the significance of the denitrification process as does the turning under of crop residues (9).

The direct evaluation of slurry as a nitrogen source accelerating the decomposition rate of straw residues is difficult considering that the chemical fractionation of organic matter is a delicate process with no practical application so far.

Theoretical results reported by Dewilde (10) from experiments carried out in the open air show a higher decomposition rate for slurry - as compared to $NaNO_3$ - sprinkled straw. This can be explained by a possible microbial priming effect of the slurry. The obvious inferiority of treatment II with regard to treatment I can only be explained by an increased immobilization of soil N when slurry is applied to straw. Finally, let us mention that the residual value, if any of slurry N never exceeds 20 kg N ha^{-1} for large dressings only (1).

2.4. Conclusions

The efficiency of slurry N is markedly related to the spreading schedule : the shorter the time interval between spreading and uptake by the crop, the higher its efficiency. Consequently, the efficiency of slurry is rather low when applied to winter cereals and larger dressings of slurry would be of no avail. An extra application of mineral fertilizer N is needed to reach the optimum yield with winter wheat and barley.

3. Influence of repeated spreadings of slurry on the principal chemical, physical and biological soil characteristics

3.1. Preliminary remarks

The amounts of major plant nutrients in the slurries vary obviously according to their nature and origin (3). Table V shows the annual fertilizer dressings and the average crop needs next to the annual amounts P_2O_5, K_2O, CaO and MgO brought with the slurry in treatments III and IV.

TABLE IV - Nitrogen effect and Efficiency of slurry.

Year	Crop	Treatment	Total N applied kg ha	Nitrogen effect kg N ha	Efficiency %
1975	Oats	III	95	-	-
		IV	127	-	-
1976	Winter Barley	III	100	-	-
		IV	160	60	37.5
1977	Sugar Beet	III	134	-	-
		IV	203	140	69.0
1978	Winter Wheat	III	122	55	46.7
		IV	174	68	39.1
1979	Winter Barley	III	137	-	-
		IV	172	55	32.0
1980	Sugar Beet	III	197	90	45.7
		IV	282	116	41.0
1981	Winter Wheat	III	126	-	-
		IV	185	65	35.1
1982	Winter Barley	III	128	-	-
		IV	203	-	-
1983	Sugar Beet	III	268	160	59.7
		IV	413	145	35.1

Next to these nutrients, slurries supply non-negligeable amounts of organic matter having a possible beneficial action on soil structure.

The last soil analyses of the plough layer have been carried out in the fall on 1982 after the harvest of the 7th crop of an 8-year crop sequence including 6 yearly applications of cattle slurry followed by applications of pig slurry as indicated in table II.

The determination of exchangeable P, K, Ca and Mg has been made following the method of Egner et al. (11).

3.2. Evolution of soil P content

Figure II shows the evolution of exchangeable P between 1975 and 1982. No statistically significant difference has been found so far between the four treatments. However, there is a slight increase as far as treatments III and IV are concerned, which results probably from the relatively high P content of pig slurry as compared to cattle slurry (table II).

The fractionation of soil P according to the method proposed by Chang and Jackson (12) underlines the positive effect of slurry application on the soluble and readily exchangeable soil P (Ofraction) as indicated in table VI. This can be explained by a reduced reversion of the organic slurry P and by the chelating action of the organic matter in the slurry (5).

3.3. Evolution of soil K content

Heavy annual dressings of cattle slurry lead very soon to a significant increase in exchangeable K in the soil. (Figure III) crop needs are relatively moderate making allowance for crop residues turned under; on the other hand K applications as fertilizer and slurry K makes K budget highly positive. The reduction or even the discontinuance of K dressings would prevent any waste in soils already well provided with this nutrient.

3.4. Evolution of soil Ca content

The amounts of Ca corresponding to slurry dressings are not negligible (table V); however they are rather small compared to the annual losses through leaching (up to 600 kg CaO ha^{-1}), (13), or to a medium liming (2,000 kg $CaCO_3$ ha^{-1} applied in 1980 on the experimental plots). On the other hand, the exchangeable Ca content is high in the soil under investigation, namely $\pm$ 500 mg Ca per 100 g of dried soil. So, no significant differences could be expected between the 4 treatments (Figure IV).

3.5. Evolution of soil Mg content

The amounts of Mg in the slurry, though rather small, are sufficient to meet the requirements of the crops (Table V). The increase in exchangeable Mg in the plough layer is shown in figure V. The effect of slurry spreadings exceeds even that of a 2,000 kg ha^{-1} application of a 55-40 calcareous magnesium amendment. Mg from livestock effluents seems to stay in or to move quickly into the soil solution and can thus be considered as beneficient to the plant (14).

TABLE V - Mean annual dressings and requirements of
investigated crops (in kg P_2O_5, K_2O, Cao, MgO ha^{-1}).

Dressings		P_2O_5	K_2O	CaO	MgO
	Mineral fertilization	60	116	1980 : 2.000	(55-40)
	Slurry Tr. III	34	155	90	35
	Tr. IV	60	229	130	50
Requirements		58	86	21	28

TABLE VI - Fractionation of soil P according to Chang
and Jackson (mg/100 g dried soil).

Fractions	Tr. I	Tr. II	Tr. III	Tr. IV
o	5.54[a]	5.87[a]	6.57[b]	7.24[b]
P(Al)	11.4	12.8	12.5	12.8
P(Fe)	15.2	15.8	15.6	15.0
P(Ca)	18.0	18.1	16.6	18.8
Total Mineral P	50.1	52.6	51.3	53.8
Organic P	27.1	28.6	30.8	28.8
Total P	77.2	81.4	82.1	82.6

Values followed by the same letter are not significantly
different at the 0.05 probability level.
The other between-treatment values are not statistically
different.

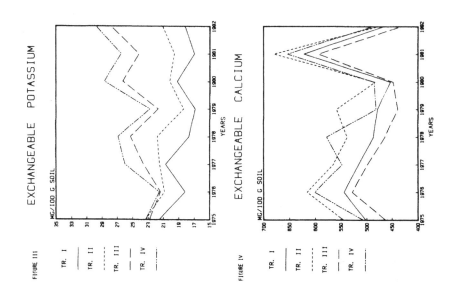

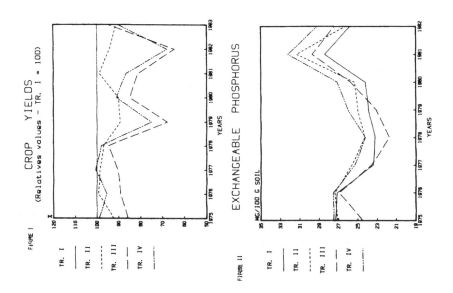

3.6. Evolution of soil pH

Slurry applications have no significant influence on soil pH (Figure VI).

3.7. Evolution of soil organic matter

Slurry applications have no influence neither on the humus content (Figure VII), nor on the organic N content (Figure VII), nor on the C/N ratio (Figure IX) of the soil.

Organic matter applied with the slurry averages only 2,200 and 3,270 kg ha^{-1} for treatment III and IV respectively. These amounts are obviously too small to have any short-term influence on the organic matter content of the soil. One should remember that repeated heavy applications of FYM - namely 40 T ha^{-1} corresponding to 10 T organic matter - show only a slight and deferred influence on soil organic matter content (15).

3.8. Evolution of soil structure

The percentage of stable aggregates and their mean-weight diameter calculated for wet-sieved soil samples have been recorded annualy during the 1978-81 experimental period. The data as represented in figures X and XI show the beneficial effect of slurry applications on soil structure.

Moreover field observations have shown, that the experimental plots treatments III and IV dried out more readily after heavy rainfalls. It may be assumed that the slurry, or the transient organic molecules formed during its decay, have a cohesion effect on the soil (16). Other authors (17), (18), (19), suppose that the increased microbial life resulting from slurry applications leads to the production of polysaccharide molecules and other extracellular polymers with an aggregating action on soil colloids.

3.9. Evolution of weeds

The modified environmental conditions resulting from repeated slurry applications as well as the weak competition of winter cereals with a low seed density have lead to the proliferation of weeds of the genera Cirsium and Tussilago setting a few problems of weed control.

3.10. Conclusions

Slurry application have a positive effect on the level of plant nutrients in the soil. In soils normally provided with these elements, the level of mineral ferti-lizing should be adapted to the amount and nature of the slurry applied. On the other hand, slurries seem to have a beneficial effect on soil structure even if they fall short of providing the adequate amount of organic matter.

4. General conclusions

The agricultural value of slurry implies the evaluation of its nitrogen efficiency on the one hand, the assessment of the influence of repeateated application of slurry on the physical, chemical and biological soil characteristics on the other hand.

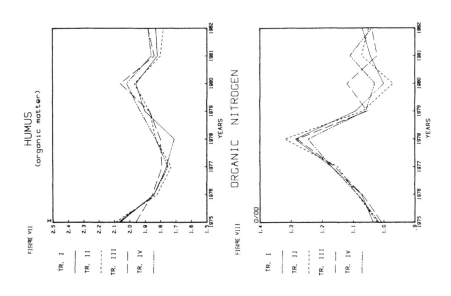

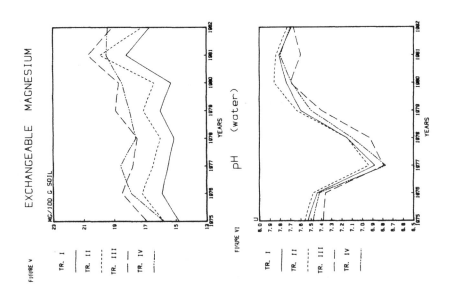

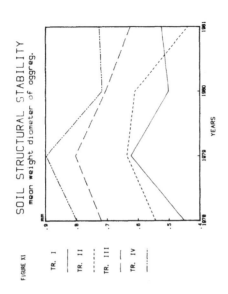

C/N RATIO

SOIL STRUCTURAL STABILITY
mean weight diameter of aggreg.

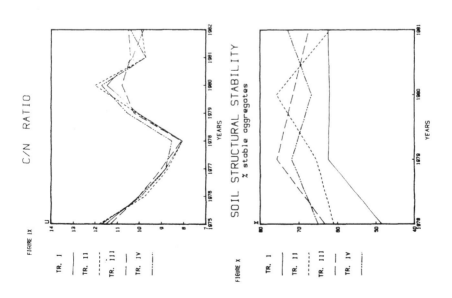

SOIL STRUCTURAL STABILITY
% stable aggregates

1. Efficiency of slurry N

It is difficult to foresee this efficiency as it depends on the climatic conditions of the cropping season and can only be calculated a posteriori. Anyhow this value is generally low especially with winter spreadings. As no effect has been observed the second year, the efficiency of slurry N is the same when calculated from annual or from long-term experiments. As slurry N alone cannot meet the plant requirements, an extra mineral N dressing should be applied especially to winter cereals.

2. Influence of slurry on the physical, chemical and biological soil characteristics

As a whole, the results of this experiment are rather positive. The increase in plant nutrients in soils already well provided will lead to the adjustment of the fertilizer dressings and consequently to important financial savings. On the other hand, the soil structure of the experimental plots seems to be improved by the slurry dressings investigated.

Acknowledgments

The authors wish to thank MM. G. Droeven and L. Rixhon heads of the Station de Chimie et de Physique agricoles and of the Station de Phytotechnie of Gembloux respectively for useful advice during the investigations. The help of A. Riga in translating and discussing the manuscript is acknowledged. Thanks are also due to Miss C. Langelez who made the typing.

References

1 - LECOMTE, R. (1980). The influence of Agronomic application of slurry on the yield and composition of Arable crop and Grassland and on changes in soil properties. Effluents from Livestock. London. Gasser, 139-183.
2 - DARCHEVILLE, M. (1973). Utilise-t-on des fumures minérales trop élevées en région limoneuse ? Revue de l'Agriculture, 5, 1091-1120.
3 - DESTAIN, J.P. et RAIMOND, Y. (1983). La composition chimique du lisier, ses facteurs de variation et ses conséquences agronomiques. Revue de l'Agriculture, 1, 36, 39-49.
4 - VETTER, H. and STEFFENS, G. (1977). The effect of nitrogen in pig slurry, spread out at different application times. Modeling nitrogen from farm wastes. London. Gasser, 44-61.
5 - VETTER, H. and STEFFENS, G. (1980). Influence of different slurry dressing on the yield and quality of plants and the nutrient contents of the shallow groundwater and of the soil. Effluents from Livestock. London, Gasser, 219-240.
5'- VETTER, H. and STEFFENS, G. (1980). Phosphorus accumulation in soil profiles and phosphorus losses after the application of animal manures. Phosphorus in sewage sludges and animal waste slurries. Haren. Hucker and Gatroux, 309-327.
6 - KOLENBRANDER, G.J. (1972). Does leaching of fertilizers affect the quality of ground water at the waterworks ? Stikstaf, 15, 8-15.

7 - GUIOT, J. (1973). La migration de l'azote dans le sol.
In "Semaine d'étude : Sol et Fertilisation". Bull. Rech.
Agron. Gembloux, Hors série, 378-385.

8 - KOLENBRANDER, G.J. (1981). Limits to the spreading of
animal excrement on agricultural land. Nitrogen losses and
surface run off. The Hague Brogan, 443-468.

9 - GANRY, F.; GUIRAUD, G. and DOMMERGUES, Y. (1978). Effect
of straw incorporation on the yield and nitrogen balance in
the sandy-pearl millet cropping system of Senegal. Plant
and Soil, 50, 647-662.

10- DEWILDE, W. (1974). Contribution à l'étude de la valeur
fertilisante du lisier et à son action sur la décomposition
des pailles. Thèse de fin d'études. Waremme, 82 p.

11- EGNER, H.; RIEHM, H. and DOMINGO, W.R. (1960). Chemische
Extraktions-methoden zur phosphor-und kalumbestimmung.
K. Lantbrukshögsk. Ann. 26, 199-215.

12- CHANG, S.C. and JACKSON, M.L. (1957). Fractionation of
soil phosphorus. Soil Sci., 84, 133-144.

13- COPPENET, M. (1974). Le problème du chaulage à la
lumière de la science agronomique moderne. Tiré-à-part dif-
fusé par le comité d'étude et de liaison des amendements
calcaires. Paris, 32 p.

14- COTTENIE, A. and VAN DE MAELE, F. (1977). Soil water
plant relationship as influenced by intensive use of effluents
from livestock. Utilization of Manure by land spreading.
Modena, Voorburg, 225-246.

15- DROEVEN, G.; RIXHON, L.; CROHAIN, A. and RAIMOND, Y.
(1980). Long term effects of different systems of organic
matter supply on the humus content and on the structural
stability of soil with regard to the crop yields in loamy
soils. Soil Degradation seminar. Wageningen, 203-222.

16- ANDERSON, F.N. and PETERSON, G.A. (1973). Agron. J.,
651, 607-700.

17- REMIE, D.A.; TRUOG, E. and ALLEN, Q.N. (1954). Soil Sci.
Soc. Am. Proc., 18, 399-403.

18- ELAINE HENNESS, J.P.; LYNCH and MAKILYN M. FLETCHER (1982).
Agregate stabilization by micro-organisms. Letcombe Annual
Report 32-33.

19- LAMAYE, J.C. (1982). Contribution à l'étude de l'acti-
vité biologique des sols par le dosage de l'A.T.P. Thèse de
Doctorat. Mons, 282 p.

EXPERIMENTS ON THE FERTILISER VALUE OF ANIMAL WASTE SLURRIES

By K A SMITH[+], *R J UNWIN and J H WILLIAMS[+]

*ADAS Soil Science Dept, MAFF, Burghill Road, Bristol BS10 6NJ, UK
[+]ADAS Soil Science Dept, MAFF, Woodthorne, Wolverhampton WV6 8TQ, UK

SUMMARY

A review is presented of the fertiliser value of animal waste slurries
according to experimental work in the United Kingdom. Results of studies
undertaken by MAFF staff since 1962, are compared with the results of
other workers. Agronomic experiments have demonstrated variable growth
responses, particularly of grass to nitrogen applied in cattle slurry.
Results suggest that frequent doses of dilute slurry are more effective
than single dressings of high dry matter material. Residual effects
have generally been small, other than following very high dressings of
slurry. Attempts to explain the variability in terms of factors such
as slurry analysis, soil type and climatic effects have met with little
success. Further agronomic experiments of the type described seem
unlikely to contribute much more to existing knowledge. Attempts at
optimising the financial return from slurries take full account of
their nutrient content and often aim to use the materials mainly as
sources of P and K.

1. INTRODUCTION

The role of organic manures in the nutrition of agricultural crops in
the UK has declined very markedly during the last 50 years. This has
occurred despite increases in livestock numbers and reflects a large
increase in the use of inorganic fertilisers and the decline of mixed
farming systems.

These trends have resulted in many farms with no access to organic
manures, whereas other holdings produce plant nutrients in animal excreta
far in excess of crop nutrient requirements. Together with intensification
has come a change on many farms from solid to liquid manure systems - from
FYM to slurry. Fifteen years ago, when environmental aspects were less
important and fertilisers were cheap, farmers and advisers were concerned
only with the disposal of slurry.

Attitudes have now changed and the emphasis once again must be on
efficient utilisation of this resource and the incorporation of manures
into an overall fertiliser policy.

The quantities of animal manures produced are considerable; total
annual UK output was estimated recently at over 160 million tonnes (21).
This represents a maximum potential fertiliser value of over £550 million
but a more realistic figure of £170 million has been suggested as the
potential value for those wastes requiring handling and return to the
land.

2. AGRONOMIC EXPERIMENTS

Chemical analyses measure the total quantities of nitrogen, phosphorus and potassium contained in organic manures but give no indication of their effectiveness or "availability" to agricultural crops. This can only be assessed satisfactorily in field experiments.

Field experiments to determine the fertiliser value of animal slurries have been undertaken for 25 years. That they still continue reflects the variable results obtained which in turn arise from the variable nature of the materials tested and the conditions under which the experiments were conducted. Generally, crop yield responses and nutrient uptake have been studied where inorganic fertilisers have been partially or completely replaced by slurries.

2.1 Efficiency & Uptake of Slurry Nitrogen

2.1.1 Experiments on Grass

Much of the work has focussed on cattle slurry and relatively few experiments have been undertaken with pig or poultry slurry. Efficiencies as high as 87% have been reported for cattle slurry nitrogen (6)(7). However, results have been extremely variable and soil type, weather conditions, slurry dilution and frequency and time of application, have all been shown to affect growth responses.

The effect of time of application has been investigated by a number of workers. Early work by MAFF Soil Scientists showed reduced effectiveness from cattle slurry applications made over winter (4). Nitrogen was about 50% effective for grass growth from February/March applications, about 25% from January and only 15% from November/December applications. A similar pattern was obtained in work at the Hannah Research Institute but more recently, trials in N Ireland have given markedly different results (7)(9). In these experiments single low rate applications of slurry in November, January or March produced efficiencies of up to 86%, with only slightly lower efficiencies for the earlier applications.

Dilution of cattle and pig slurry has increased efficiency and several small dressings have been shown to be more effective than a single application (23).

Similar effects have been reported in work at Bridgets Experimental Husbandry Farm, with a 90% effectiveness of dilute (2-4% dry matter) slurry applied on 10 occasions to give 300 kg/ha N (2). Effectiveness declined rapidly as total N rates increased above crop requirements and a large seasonal variation in efficiency was observed over the 4 years of the trials. (Table I).

Table I - Nitrogen efficiency from repeated winter application of
cattle slurry (2).

Slurry N applied Kg/ha	Mean effectiveness compared to NH_4NO_3 (%)	Range of effectiveness according to season (%)
300	90	65-131
600	63	45-89
1200	38	27-53

Unwin found in a three year trial in SW England, that the N in thick
slurry (10.5% dry matter) was 60% effective when spread in six monthly
applications each winter and compared with ammonium nitrate applied during
the growing season (24).

Advisory literature in the 1970s suggested a 50% availability for
cattle slurry N from a late winter/early spring application. Work
undertaken since that time however, has tended to suggest a rather lower
availability. Pain & Saunders calculated different efficiencies for each
cut of grass; whilst a March slurry dressing was 50% effective, the June
application was only 15% (19). Averaged over the season, slurry N was only
25-30% as efficient as inorganic N.

A series of seven trials were undertaken on 4 sites in S England in
1977-79, with slurry applied at 3 times to ryegrass in factorial
combination with ammonium nitrate (25). Results were variable but average
effectiveness of slurry N in the absence of NH_4NO_3 was 25-30% for
spring and summer applications and 15% for winter dressings.

Subsequently a national series of 27 trials were undertaken by MAFF on
sites throughout England and Wales in 1980-82(22). Target rates of slurry,
supplying nil, 80 and 160 kg/ha N were applied in early spring or after the
first cut and compared with NH_4NO_3. Slurries were typically around
7-8% dry matter and 0.3% total N content, with low and high rates of
application of roughly 25m^3/ha and 50m^3/ha respectively. Application
rates, however, ranged up to 134 m^3/ha for the more dilute (eg 1.6% dry
matter content) slurries.

Effectiveness of slurry N was estimated using fitted quadratic curves
for the inorganic N response and the mean values are shown in Table II for
the first cut after slurry application.

Table II - Slurry nitrogen efficiency (%) in a National Series of
Trials (1980-82).

| Year | Early Spring (1st cut) | | Late application (2nd cut) | |
	Low rate	High rate	Low rate	High rate
1980	34 (8 sites)	24 (8 sites)	21 (7 sites)	18 (7 sites)
1981	42 (9 sites)	24 (9 sites)	29 (8 sites)	17 (9 sites)
1982	37 (7 sites)	21 (8 sites)	21 (5 sites)	19 (6 sites)
Mean	38	23	24	17

NB: At some sites cuts were lost due to farming activities.

Site location and rainfall following application had no consistent
effect on responses. Slurry N efficiency appeared to be slightly higher on
sites with predominantly silty topsoils as opposed to sites on soils of
higher clay content. In contrast to authors who have shown a close
correlation between slurry N efficiency and "soluble" nitrogen content, the
relationship between NH_4-N content and efficiency was poor (14). The low
efficiency of the higher rate of application was likely to be at least in
part due to physical smothering of the sward observed at some sites. The
results also indicated an additive effect of slurry N and inorganic N, when
both sources of N were applied together.

2.1.2 Experiments on Arable Crops

Crop yields in experiments conducted in N Ireland have shown a good
correlation between responses to NH_4-N in slurry and fertiliser N in the
year of application (12). Soluble nitrogen content (extraction with cold
0.1N HCl) was found to be at least as effective as urea-N and ammonium
sulphate-N when applied to spring barley seedbeds.
MAFF experiments (1965-67) on the fertiliser value of pig slurry for
spring barley were carried out in E Yorks (5). Values for the efficiency
of the slurry N, estimated from the inorganic N response curve, varied
widely (Table III).

Table III - Range of pig slurry N efficiency for barley (5).

| Rate of slurry N (kg/ha) | Time of Application | |
	Winter	Spring
≤190	35%	57 - 105%
190-380	18%	27 - 36%

It was suggested that when applied at up to 190 kg/ha total N to the seedbed, 50-75% is available; at higher rates 25-50% is available. About 25% of the nitrogen was considered available from mid-winter slurry applications.

2.1.3 Residual Effects

The residual effects of large/repeated dressings of solid manures are well known and have been shown to persist for many years, for example in work at Rothamsted and Woburn (10). Slurry experiments, however, have usually been undertaken for one year only. Where experiments have continued beyond the first year, the results have indicated rather small residual effects, except where very high dressings were applied.

Some residual effects were observed in MAFF grass experiments with poultry (battery) manure (27). Autumn or spring applied manure supplying roughly 88 and 176 kg/ha N, was compared with fertiliser N applied in autumn, spring or in split dressings between spring and early summer. When applied in August, dressings of manure or fertiliser produced an immediate response, with poultry manure N about 66% as effective as fertiliser N; applied later in the autumn, manure or fertiliser N gave only small residual effects the following spring. At one site the efficiency of spring applied poultry manure N increased from 50% to about 80% in the second and third year of the experiment, suggesting an appreciable residual effect or carry-over of manure N from previous years.

Cromack described studies into factors limiting the application of slurry on grass and arable land (3). Dilute cow slurry was applied through an organic irrigation system, overwinter in frequent dressings, to free draining silty loam soils over chalk, at Bridgets EHF. Potatoes were grown in the first year and the residual value of the slurry to winter wheat grown the following year is shown in Table IV. The response to N top dressing declined with increased slurry residues.

Table IV - Residual effects of cattle slurry on winter wheat. (Yield of wheat at 85% DM, t/ha).

Slurry applied (m^3/ha)		Control	595	1255	1860
Total slurry nitrogen (kg/ha)		–	850	1820	2680
Rate of mineral N - Nil		2.4	3.7	5.0	5.9
Rate of mineral N - 50 kg/ha		3.7	5.2	5.7	5.2
Rate of mineral N - 100 kg/ha		4.3	5.6	5.4	4.4

In NIRD/MAFF experiments, the effects of cattle slurry on spring barley were investigated over 3 years at NIRD (18). Large residual benefits were observed from cow slurry applied at rates of up to 112 t/ha before a barley crop in the previous year. Satisfactory grain yields were obtained in the second year without additional fertiliser.

In MAFF/NIRD experiments on grass, single applications of cow slurry containing 200-420 kg/ha N produced small residual yield effects in the first two cuts the following year (19)(25). The uptake of slurry N was 3-10% of that applied. In 3 out of 4 experiments tested, a measurable increase in yield was also found from 300 kg/ha N applied as NH_4NO_3 the previous year.

In Northern Ireland, Adams reported similar responses in the first and second residual years of experiments, with no evidence that high N use plus slurry resulted in N accumulation in the soil (1). Similarly in long term trials with pig and cow slurry, similar efficiencies of slurry N have been recorded over many years of application to the same plots, indicating little residual effect (14).

In experiments on the use of cow slurry for maize, Pain & Saunders reported an apparent recovery of slurry N of 17%. Slurry application in 1977, supplying up to 276 kg/ha N, however, produced no measurable residual effects when tested with a second crop of maize, in 1978(19).

In the recent ADAS national series of trials on the fertiliser value of cattle slurry N, residual effects in the season of application were assessed by taking additional grass cuts from slurry treated plots without further addition of slurry or mineral N(23).

The largest effects, as expected, were observed following the late slurry treatment. (Table V).

Table V - Residual (late season) effects of slurry N for grass

Efficiency in terms of dry matter response (% of original slurry N applied)

	EARLY APPLICATION*				LATE APPLICATION*	
	Cut 2		Cut 3		Cut 3	
Slurry rate**	Low	High	Low	High	Low	High
1980	5	5 (8)	2	3 (6)	8	8 (7)
1981	5	3 (9)	5	3 (9)	8	6 (9)
1982	4	4 (7)	3	2 (6)	6	5 (6)
Overall Mean	5	4	4	3	7	6

* Early - March/April; Late - after first cut (May/June)
** Target slurry N rates - low 80 kg/ha N; high 160 kg/ha N
( ) No of sites

In this study, clover proliferation on some of the slurry plots and lower mineral N plots may have resulted in a slight overestimation of "residual" effects.

Recent MAFF experiments in Lincolnshire have attempted to assess the fertiliser value of winter applied pig slurry nitrogen for spring barley (11). Slurry supplying 0, 100, 200 and 300 kg/ha N was applied in winter,in factorial combination with fertiliser N top dressed after crop emergence.

The residual value of slurry N for the spring barley varied with season but was generally low (Table VI), even at low rates of fertiliser N application.

Table VI - Efficiency of winter applied pig slurry N for spring barley.

Season	Date of slurry application	
1980/81	Dec 4	10%
1981/82	Jan 14	20-25%
1982/83	Dec 21	6-20%

Because of higher winter and early spring rainfall in 1981 and 1983 the lower residual nitrogen effects in those years were expected.

Thus, in addition to the low immediate availability and low apparent herbage recovery of slurry nitrogen recorded in trials, residual responses are also often low. One reason may be the proportion of nitrogen remaining in the surface soil; the additional N from the slurry is often only a small proportion of total N in the soil. In trials on MAFF Experimental Husbandry Farms, after 4 annual treatments with pig slurry, only 4% of a total application of 14,000 kg/ha N of slurry could be detected by analysis in the top 15 cm of soil, whilst 15% of a 2500 kg/ha N rate was recovered (8). Similar trials with cow slurry on different soils suggested an 18-30% retention of slurry N with no consistent effect of rate of application in the range 1250-8,000 kg/ha N. Unwin also found a greater increase in total soil N after applications of cow slurry than after pig slurry (Table VII)(24). In this trial, some plots received slurry intermittently over an 18 month period before the plots were reseeded in April 1974. The results give a clear indication that the residual effect of the slurry N was small or short-lived.

2.2 Phosphorus and Potassium - Effects on Soils and Herbage

Responses to slurry phosphorus and potassium are more difficult to assess than those to nitrogen. Even if responsive sites are available for testing other factors confound the results and workers can usually only assess availability in terms of uptake by herbage.

A 50% availability for P is often assumed and trials in N Ireland have tended to confirm this (23). However, on some less responsive sites, there has been little difference between fertiliser and slurry P(13). Other results indicating much lower short term recoveries of P were thought to be due to the organic binding of much of the P in slurry (1).

In some MAFF slurry experiments phosphorus levels in herbage were increased to a similar degree by cow or by pig slurry although the absolute effects varied with soil type (26). Two years after applications ceased at one site, P values returned to control levels even though up to 1500 kg/ha P had been applied. (Table VIII). The proportion of the P balance which could be detected in the soil as sodium bicarbonate extractable varied with type of slurry and rate of application. At low rates of pig slurry up to 40% could be accounted for in this form but only 10% at higher rates.

Table VII - Residual yield effects, nitrogen recovery and soil nitrogen content after sacrificial dressings of pig and cow slurry to a heavy soil.

	Total Slurry N applied kg/ha	Dry Matter Yield kg/ha			Estimated Efficiency of Slurry N			% Recovery of Applied N			% Total Soil N 0-15 cm
		1974	1975	1976	1974	1975	1976	1974	1975	1976	1976
Cow Slurry	3028	12409	8888	4820	26*	9	3	13	8	0.5	0.33
	6056	14047	11310	5114	17*	7*	1	6	4	0.2	0.37
Pig Slurry	1920	12774	7042	3001	44*	9	0	16	5	0	0.29
	4446	14995	9366	2724	25*	7	0	9	4	0	0.27
	14780	14505	14349	4545	7*	4*	1	3	3	0.3	0.30
Control Nil Slurry Nil NH_4NO_3		5743	3702	2528	-	-	-	-	-	-	0.27

* Best estimate. Yield greater than from NH_4NO_3 controls.

Small increases in extractable P were noted at depth in the soil, although the majority is retained within 30 cm of the surface. Recovery of cow slurry P by soil analysis was always less than 13% which is in agreement with the 15% quoted by workers in N Ireland (14). When heavy rainfall follows slurry application, P may be translocated through soils both in soluble and particulate forms associated with movement of slurry through the soil pores (15).

Table VIII - Effect of cattle slurry phosphorus on herbage analysis

Total slurry m^3/ha	Nil	292	584	810	1620
Total P kg/ha	189*	246	492	757	1514
% P in herbage DM	0.39	0.40	0.39	0.39	0.50
% P in herbage 2 years after slurry application ceased.	0.28	0.26	0.24	0.26	0.29

* inorganic fertiliser

Slurry K has been found to be almost equivalent to fertiliser K by some authors but rather less so by others. The K fixing properties of different soils will obviously affect response to this nutrient. Applications of pig and cow slurry have been shown to raise available soil K levels less on a silty clay loam than on silty loams (20). On a silty clay loam soil, 9250 kg/ha K in cow slurry produced very large increases in available soil K and herbage K (Table IX)(24). After two years with no further treatment, the soil K had fallen appreciably and herbage K values were lower than on plots receiving regular fertiliser dressings.

Table IX - Effects of high rates of cow slurry K on herbage & soil analysis

	Total K applied 1972-1976 (kg/ha)	Herbage analysis 1974 (% K in DM)	1976	Soil analysis 1974 (available K mg/l)	1976*
NPK Fertiliser	81	1.6	1.5	57	80
	475	2.1	2.0	64	85
	874	2.4	2.4	63	125
Slurry	4623	4.4	2.3	500	280
	9246	5.1	2.0	847	450

* 5-15cm depth soil sample.

Available soil K values can increase at depth following prolonged slurry use. A two fold increase in available K throughout the top 75 cm of a fine sandy loam soil has been recorded following applications of pig and cow slurry at sacrificial rates over a period of 15 years (20).

3. THE VALUE OF SLURRIES IN FARM PRACTICE

The results discussed have been incorporated into MAFF advice to farmers in England and Wales (16). The many variables which influence the effectiveness of slurry nutrients means that such guidance has to be sensibly interpreted for an individual farm situation. The proportion of total nutrients that are assumed to be available in the season of application is reproduced in Table X. Nitrogen efficiency needs to be further modified according to time of application.

Table X - Proportion of total nutrients available in animal waste slurries in the season of application.

Type of slurry	N(a)	P_2O_5	K_2O
		% available	
Cattle	30(b)	50	90
Pig	65	50	90
Poultry	65	50	90

(a) The N figures apply to manures and slurries spread in spring. In autumn and winter these proportions will be smaller.

(b) For slurry applied as a surface dressing to grassland. When incorporated into the soil during or soon after spreading the N availability may be as high as 50%.

Nitrogen supplementation with inorganic fertiliser can be determined according to crop nutrient requirements, the guidelines for manure N availability and local experience. Such an approach is considered to be the most satisfactory way of optimising the fertiliser value of manures. Regular soil analysis is advised when a large proportion of the total P and K requirements are supplied by organic manures.

Much of the non available slurry phosphorus is in the form of complex inorganic and organic compounds. The slow decomposition of these compounds in the soil will release soluble phosphate which is available for plant uptake. Over a period of years, therefore, it may be appropriate to consider a much higher efficiency of slurry P in a maintenance manuring policy. The more readily available slurry potash will tend to be removed by luxury uptake if over-applied in one year and residual effects are likely to be less evident.

In practice, the nitrogen value of animal waste slurries is likely to be signficantly reduced by gaseous and leaching losses following application to the soil depending on soil and weather conditions (22).

The residual value of slurry nitrogen has been taken into account in MAFF fertiliser recommendations, according to the soil nitrogen index (17). However, it is likely that the index system will require further adjustment in view of the minimal residual effects observed from other than high rates of slurry application.

When attempts are made to maximise the fertiliser value of organic manures and slurries, the extra costs incurred often outweigh the benefits obtained. When capital costs for various slurry storage systems are combined with collection and spreading costs and expressed as an annual charge, it can be shown that slurry storage and handling may cost in excess of its nutrient value.

However, many farmers are prepared, or have to accept such costs, either because they provide a more manageable system or because they are dictated by environmental considerations.

By not attempting to spread slurry throughout the winter, farmers can avoid damage to soils and grass swards. By seeking to optimise fertiliser recovery, the risks of damaging soil and causing pollution are usually minimised.

4. CONCLUSIONS

Animal waste slurries are a valuable resource which is not being fully exploited at the present time.

Responses to slurry nutrients, particularly nitrogen, have been extremely variable and often disappointing. Where residual effects have been recorded, these have usually been small, except where high dressings of slurry have been applied to preceeding crops. Despite considerable experimental effort, our understanding of the factors which contribute to the net growth response of crops, particularly grassland, remains inadequate. Further agronomic experiments of the type described are perhaps unlikely to contribute much more to existing knowledge.

Some of the practices which could be adopted to increase the efficiency of the nitrogen response would cost the farmer much more than the value of the benefit obtained. However, these practices may have to be considered further in view of increasing environmental constraints.

REFERENCES

1. ADAMS S N (1974). The responses of pastures in Northern Ireland to N, P and K fertilisers and to animal slurries.
   iii. Effects in experiments continued for either two or three years.
   J Agric Sci Camb 82 129-137.
2. APPLETON M, RICHARDSON S J (1976). Cow slurry management with particular reference to Bridget's. ADAS Q Rev 23 294-305.
3. CROMACK H T H (1971). The disposal of cow slurry on land.
   Bridgets EHF, Annual Report No 12, MAFF.
4. DAVIES H T (1970). Experiments on the fertilising value of animal slurries.
   Part 1 - The use of cow slurry on grassland.
   Experimental Husbandry 19 49-60.
5. DAVIES H T (1970). Experiments on the fertilising value of animal slurries.
   Part II - The use of pig slurry on spring barley.
   Experimental Husbandry 19 61-64
6. DRYSDALE A D (1963). Liquid manure as a grassland fertiliser.
   ii. The response to winter applications. J Agric Sci Camb 61 353-360
7. DRYSDALE A D (1965). Liquid manure as a grassland fertiliser.
   iii. The effect of liquid manure on the yield and botanical composition of pasture and its interaction with nitrogen, phosphate and potash fertilisers.
   J Agric Sci Camb 65 333-340.
8. GOSTICK K G (1981). Accumulation of organic nitrogen in soils after application to grassland. Some studies in England and Wales. In Nitrogen losses and surface runoff from landspread manures.
   Ed J C Brogan, Nijhoff/Junk CEC.

9. GRACEY H I (1983). Efficiency of slurry nitrogen as affected by the time and rate of slurry application and rate of inorganic nitrogen. In Efficient Grassland Farming. Proceedings of European Grassland Federation Conference, Reading 1982.

10. JOHNSON A E (1970). The value of residues from long-period manuring at Rothamsted & Woburn II. A summary of the results of experiments started by Lawes & Gilbert.
Rothamsted Report for 1969 Part 2. 7-21.

11. JOHNSON P A & PRINCE J (1983). Fertiliser value of pig slurry nitrogen for spring barley. Internal MAFF reports.

12. MCALLISTER J S V (1970). Relative efficiencies of different nitrogen sources for barley in N Ireland.
Agric Res Institute of NI. 43rd Annual Report 1969-70 23.

13. MCALLISTER J S V (1977). Efficient recycling of nutrients. In Utilisation of manure by land spreading.
Ed J H Vorburg 87-105 CEC.

14. MCALLISTER J S V (1981). Responses to slurry N in Northern Ireland. In Nitrogen losses and surface run-off from land spread manures. Ed J C Brogan 389-393. Nijhoff/Junk CEC.

15. MCALLISTER J S V, Stevens R J (1981). Problems related to phosphorus in the disposal of slurry. In Phosphorus in sewage sludge and animal waste slurries.
Ed T W G Hucker and G Catroux. 383-396. Reidel. Dordrecht.

16. MAFF (1982) Profitable utilisation of livestock manures. Booklet 2081, MAFF London

17. MAFF (1983). Fertiliser recommendations
Reference Book 209, Ministry of Agriculture, Fisheries and Food, London.

18. PAIN B F, RICHARDSON S J, Fulford R J (1978). The effects of cattle slurry inorganic nitrogen fertiliser on the yield and quality of spring barley. J Agric Sci Camb 90 283-289.

19. PAIN B F, SAUNDERS L T (1980). Effluents from intensive livestock units fertiliser equivalent of cattle slurry for grass and forage maize. In Effluents from Livestock. Ed J K R Gasser. 300-311 Applied Science Publications Ltd.

20. RUSSELL R D, UNWIN R J (1982). Potassium in soils and herbage of intensive livestock farms. Internal MAFF report.

21. SMITH K A & UNWIN R J (1983). Fertiliser value of organic manures in the UK. Fertiliser Society Proceedings No. 221.

22. SMITH K A, PAIN B F & DYER C J. Fertiliser equivalent of nitrogen in cow slurry for conserved grass; ADAS Trials 1980-82. (Report in preparation).

23. STEWART T A (1969). The effect of age, dilution and rate of application of cow and pig slurry on grass production. Record of Agric Res 17 (1) 68-90. (Ministry of Ag NI).

24. UNWIN R J (1973-1976). Slurry application to grassland. In Soil Science Reports of Experiments in SW Region of MAFF 1973-1976.

25. UNWIN R J, PAIN B F and WHINHAM W N. The effect of rate and time of application of nitrogen in cow slurry on grass cut for silage. (Report in preparation).

26. UNWIN R J (1981). Phosphorus accumulation and mobility from large applications of slurry. In Phosphorus in Sewage Sludge and Animal Waste Slurries. Edited T W G Hucker, G Catroux, Reidel Dordrecht.

27. WEBBER J & BASTIMAN B (1967). Experiments testing poultry manure as a source of nitrogen for grass. Experimental Husbandry 15 11-17.

LONG-TERM EFFECTS OF SEWAGE SLUDGE AND PIG SLURRY APPLICATIONS ON
MICRO-BIOLOGICAL AND CHEMICAL SOIL PROPERTIES IN FIELD EXPERIMENTS

F.X. STADELMANN and O.J. FURRER
Swiss Federal Research Station for Agricultural Chemistry and Hygiene
of Environment
3097 Liebefeld-Berne (Switzerland)

Summary

In long-term field experiments the following microbiological and
chemical soil changes have been observed:
1. Long-term applications of sewage sludge and pig slurry increased soil humus content, N-content, CEC, heterotrophic
   soil microorganisms and their activities more in a light
   soil than in a heavy soil.
2. In a 7 year field experiment on a sandy loam soil (parabrown
   soil) receiving annually 5 tons of organic matter per ha in
   the form of sewage sludge, appreciable increases in the humus content, N content, pH values, contents of aerobic bacteria and biological activities were observed. These changes
   were clearly seen up to a soil depth of 1 meter.
3. Applications of sewage sludge at annual rates of 4 and 15
   tons of organic matter per ha in a 5 year field experiment
   on a loamy soil used for grassland and arable crops have in-
   duced measurable aftereffects on soil fertility. In sewage
   sludge fertilized soils, 7 years after the last application
   of sludge, an appreciable increase in the humus, extractable
   P and Zn content and in microbiological activities was found
   when compared with soils treated earlier with mineral fertilizer. The aftereffects due to heavy pig slurry doses were less
   pronounced.

1. INTRODUCTION

Sewage sludge and farm manures such as pig slurry contain considerable amounts of organic matter and plant nutrients (N, P, K, Ca, Mg)
(2, 3, 5). It is known that land application of organic manures and
wastes not only increases plant production (2, 3) but also the humus
content (2, 5, 6) and the microbial activities (1, 2, 4, 5, 6) in the
surface soil layer. Sludge application usually has a greater effect on
the soil microorganisms and their activities in grassland soils than
in arable soils (6). By contrast, there is little information about
the changes in chemical and microbiological properties in a soil profile which may occur due to application of organic manure. Likewise,
the possible long-term aftereffects of organic manure on soil fertility have hardly been investigated.

Table I Influence of a 8 year application (1976-1983) of mineral fertilizer (Min), of sewage sludge(SS$_5$ = 5 t organic matter/ha · year) and of pig slurry (PS$_5$ = 5 t organic matter/ha · year)on some chemical and microbiological properties (surface layer 0-10 cm) of two different soils (field experiment Liebefeld and Büren)

Location	Fertilizer treatments	Mean (Ø 1976-1983) grass yield (t d.m./ha)	Chemical soil properties in 1983				Microbiological soil properties (relative values) in 1983								
			pH (H$_2$O)	C (%)	N (%)	CEC (meq/100 g)	Aerobic bacteria	Actinomycetes	Ratio-Bacteria/fungi (units)	Algae	N-fixing blue-green algae	Catalase activity	Alkaline phosphatase activity	C-Mineralization	N-Mineralization
Liebefeld (light soil)	0	11.8	6.0	1.50	0.243	17.2	100	100	100	100	100	100	100	100	100
	SS$_5$	14.1	6.9	2.58	0.337	23.7	300	200	146	90	86	196	243	173	88
	PS$_5$	16.1	6.0	2.03	0.291	22.2	246	146	199	126	92	226	124	85	122
Büren (heavy soil)	0	9.5	7.3	2.34	0.356	28.0	289	152	136	336	178	297	358	218	93
	SS$_5$	13.1	7.3	2.52	0.429	31.6	523	200	392	136	97	338	343	239	103
	PS$_5$	13.2	7.3	2.26	0.409	30.0	498	324	387	505	342	500	352	233	141

138

The objectives of the investigations presented in this paper were
to study:
a) the effect of long-term heavy applications of sewage sludge and pig
slurry on some chemical and microbiological soil properties depending
on soil type and soil texture (a light and a heavy soil),
b) the effect of long-term heavy applications of sewage sludge on the
chemical and microbiological soil properties in a soil profile,
c) the occurrence of possible aftereffects in chemical and microbiological soil properties after a longer break in heavy applications of
sewage sludge and pig slurry.

2. EFFECTS ON BIOCHEMICAL PROPERTIES OF TWO DIFFERENT SOILS

In a field experiment at two different locations (Liebefeld: light
soil, Büren: heavy soil), permanent grassland plots (clover grass) were
fertilized annually for 8 years (1976-1983) either without fertilizer
(0 = control) or with high amounts of sewage sludge (SS_5 = mean annual
doses including additional mineral fertilizer: 5,000 kg organic matter,
420 kg N of which 280 kg were organically bound, 380 kg P, 250 kg K,
1,040 kg Ca per ha)or with high amounts of pig slurry (PS_5 = mean annual doses: 5,000 kg organic matter, 585 kg N of which ca. 200 kg were
organically bound, 141 kg P, 230 kg K, 220 kg Ca per ha).
The two locations are characterized as follows (3):
Liebefeld: 565 m above sea level; average annual rainfall 1,141 mm;
average annual air temperature 8.0^o C ; soil type: weakly developed
parabrown soil;pH ca. 6.3; C_t ca. 2.0; soil texture sandy loam (16 %
clay, 55 % sand).
Büren a.A.: 430 m above sea level; average annual rainfall 1,326 mm;
average annual air temperature 8.9^o C; soil type: gleyed calcareous
brown soil; pH ca. 7.3; C_t ca 2.4 %; $CaCO_3$ ca. 25 %; soil texture:
silty clayey loam (42 % clay, 7 % sand).
Soil samples from the 0-10 cm layer (3 replications per treatment and location) were collected at the end of the season (21.11.1983)
and analyzed for chemical and microbiological soil properties according
to the methods described by STADELMANN (5) and STADELMANN et al. (7).
As shown in Table I,a long-term application of high doses of organic wastes and manures increased the humus content (C_t), N-content
and the cation exchange capacity (CEC) in the surface soil layer. Sewage sludge had a greater effect on these chemical soil properties than
pig slurry. Sewage sludge and pig slurry influenced the chemical soil
properties more drastically in the light soil (Liebefeld) than in the
heavy soil (Büren).
In agreement with previous observations (1, 4, 5, 6) sewage sludge
and pig slurry applications increased counts of heterotrophic microorganisms, activities of enzymes and mineralization of organic carbon and
nitrogen in the soils: Table I. In contrast to pig slurry, sewage sludge
reduced the soil algae and N-fixing blue-green algae. In accordance with
increased plant yields, pig slurry had also a more pronounced positive
effect than sewage sludge on the microbiological indicators of high soil
fertility, viz. ratio between bacterial counts and fungal counts, catalase activity and N-mineralization (5).
The heavy soil (Büren) showed higher initial contents of heterotrophic and autotrophic microorganisms and higher biological activities

Table II Influence of a 7 year application (1976-1982) of sewage sludge (SS$_5$ = 5 t organic matter/ha · year) compared to mineral fertilizer (Min) on some chemical and microbiological properties of a soil profile (field experiment Liebefeld)

Fertilizer treatments	Hori-zon	Soil depth (cm)	Chemical soil properties			Microbiological soil properties (relative values)					
			pH (H$_2$O)	C (%)	N (%)	Aerobic bacteria	Actino-mycetes	Catalase activity	Alkaline phospha-tase activity	C-Minera-lization	M-Minera-lization
Min	Ahp	0- 2	6.5	2.20	0.284	100	100	100	100	100	100
		2- 5	6.4	1.71	0.259	31	90	75	76	52	66
		5- 10	6.5	1.53	0.245	22	36	48	74	47	59
		10- 15	6.6	1.40	0.224	18	29	41	86	50	56
		15- 25	6.8	1.26	0.209	18	22	43	86	37	46
	Ast	25- 40	6.9	0.98	0.159	26	10	31	71	28	18
	Bw	40- 70	6.9	0.65	0.098	19	18	20	27	24	21
	It	70-100	7.0	0.43	0.074	20	-	3	21	16	10
SS$_5$	Ahp	0- 2	7.1	3.42	0.452	174	127	169	147	205	173
		2- 5	7.2	2.87	0.370	49	94	106	126	110	96
		5- 10	7.2	2.16	0.308	46	82	70	118	93	73
		10- 15	7.2	1.95	0.260	22	58	45	115	71	58
		15- 25	7.1	1.59	0.288	22	29	39	106	63	45
	Ast	25- 40	7.2	1.31	0.179	34	22	28	95	52	37
	Bw	40- 70	7.3	0.79	0.106	20	10	15	41	45	21
	It	70-100	7.3	0.55	0.077	25	14	5	28	25	12

than the light soil (Liebefeld). The organic manures increased the biomass of the heterotrophic microorganismus (bacteria, fungi) and their
activities (catalase, alkaline phosphatase) considerably more in the
light soil than in the heavy soil. Autotrophic algae and N-fixing blue-
green algae however were more drastically reduced by sludge application
in heavy soil than in light soil.

3. EFFECTS ON BIOCHEMICAL SOIL PROPERTIES IN A SOIL PROFILE

In the field experiment at Liebefeld (light soil), described in
the previous chapter, permanent grassland plots were yearly fertilized
either with mineral fertilizer (Min = 220 kg N, 50 kg P, 250 kg K,
212 kg Ca/ha · year) or with high amounts of sewage sludge (SS_5 =
5,000 kg organic matter/ha · year). After 7 years, soil profile samples
(pooled samples of 4 replications per treatment and horizon) were collected at the end of the season (1.11.1982) and analyzed for chemical
and microbiological soil properties according to the method described
in (5) and (6).

With increasing depth, a decrease in humus content (C_t), nitrogen
content and the number of soil microorganisms and their mineralization
and enzyme activities was found (Table II). Compared to mineral fertilizer, sewage sludge application increased the pH values (Fig. 1), the
N-content (Fig. 2), the humus content (Fig. 3), the numbers of aerobic
bacteria, the activity of soil enzymes and the mineralization of organic C (Fig. 4) and N down to a depth of 1 m. The enhanced microbial activity in the deeper soil layers can be explained by increasing supply
of organic C and N. The increased levels of total C and N are possibly
due to enhanced root excretion and residues due to sludge application.
In this process, earthworms perhaps played an important role by mixing
the surface soil with subsurface soil. The observed increase in soil pH
is due to the extra input of calcium through sewage.

4. AFTEREFFECTS ON BIOCHEMICAL SOIL PROPERTIES

In a field experiment, carried out on a heavy soil for 5 years
(1971-75), sewage sludge (SS) and pig slurry (PS) were applied at an-
nual rates of 120 m^3 (SS_4, PS_4) and 480 m^3 (SS_{15}, PS_{14}), respectively
(Table III). Applications were done on both grassland and arable soil
(3 year crop rotation: wheat - maize - sugar beet). Experimental site:
Belp, 510 m above sea level, average annual rainfall 972 mm, average
annual air temperature 8.8^o C. Soil type: alluvial, compact gley, poor
in skeleton. Soil texture: loamy soil (61 % clay, 6 % sand). High doses
of sewage sludge (SS_{15}) or pig slurry (PS_{14}) increased the yields of
grass and sugar beet but not of wheat, compared to mineral fertilizer
(Min) (2).

During 1975-81, the whole experimental area was used as grassland,
with mineral fertilizers and cow slurry being spread evenly, whereas in
1982 winter wheat was grown. On 16 August 1982, the 7th year after the
last application of sewage sludge or pig slurry, soil samples (3 replications for each treatment) from the surface soil (0-20 cm) were taken
and analyzed for chemical and microbiological properties (5, 6).

The aftereffect of organic waste application was still visible
even 7 years after the last application of sewage sludge or pig slurry

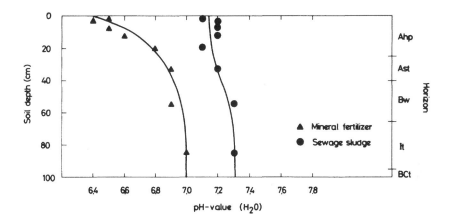

Figure 1: Influence of a 7 year application of sewage sludge (SS_5 = 5 t organic matter/ha · year) compared to mineral fertilizer (Min) on the pH-values of a soil profile (field experiment Liebe-feld).

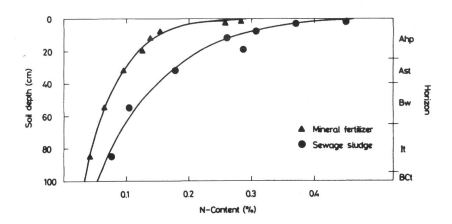

Figure 2: Influence of a 7 year application of sewage sludge (SS_5 = 5 t organic matter/ha · year) compared to mineral fertilizer (Min) on the N-content of a soil profile (field experiment Liebe-feld).

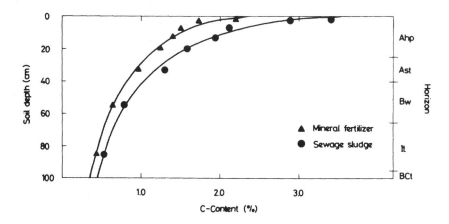

Figure 3: Influence of a 7 year application of sewage sludge (SS$_5$ = 5 t
organic matter/ha · year) compared to mineral fertilizer (Min)
on the humus content (C$_t$-content) of a soil profile (field
experiment Liebefeld).

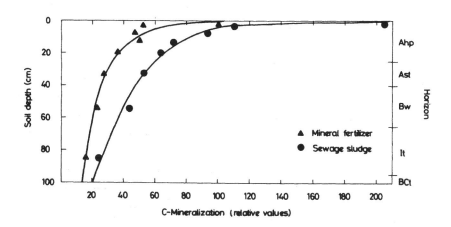

Figure 4: Influence of a 7 year application of sewage sludge (SS$_5$ = 5 t
organic matter/ha · year) compared to mineral fertilizer (Min)
on the mineralization of organic C (soil respiration) of a
soil profile (field experiment Liebefeld).

Table III Chemical and microbiological soil properties (surface layer 0-20 cm) 7 years after the last application of sewage sludge and pig slurry (field experiment Belp)

Crops during 1971-75	Fertilizer treatments 1971-75	Chemical soil properties in 1982						Microbiological soil properties in 1982 (relative values)				
		pH (H₂O)	C$_t$ (%)	N (%)	P extractable (ppm)	K extractable (ppm)	Zn (ppm)	Aerobic bacteria	Catalase activity	Alkaline phosphatase activity	C-Mineralization	N-Mineralization
Permanent grassland	Min	6.5	3.95	0.527	0.47	0.57	118	100	100	100	100	100
	SS₄	6.8	4.12	0.556	1.34	0.57	182	125	·110	109	94	102
	SS₁₅	7.1	4.21	0.563	4.58	0.63	295	138	121	106	170	114
	PS₁₄	6.3	4.34	0.550	1.32	0.53	144	81	110	99	81	110
Arable crops	Min	6.1	3.80	0.521	0.66	0.70	n.d.	76	115	82	66	92
	SS₁₅	7.0	4.07	0.553	3.85	0.65	242	121	135	96	149	110

Min = Mineral fertilizers (grassland: 275 kg N, 75 kg P, 400 kg K and 363 kg Ca/ha · year; arable land: 150 kg N, 40 kg P, 200 kg K and 194 kg Ca/ha · year)

SS₄ = 120 m³ sewage sludge (4345 kg organic matter, 483 kg N, 164 kg P, 412 kg K and 1240 kg Ca/ha · year. K supplied mainly as mineral fertilizer)

SS₁₅ = 480 m³ sewage sludge (15210 kg organic matter, 1132 kg N, 584 kg P, 425 kg K and 4400 kg Ca/ha · year. K supplied partially as mineral fertilizer)

PS₁₄ = 480 m³ pig slurry (13610 kg organic matter, 1197 kg N, 348 kg P, 526 kg K and 757 kg Ca/ha · year. K supplied partially as mineral fertilizer)

n.d. = not determined

144

(Table III). There was an appreciable increase in the contents or organic matter (C_t), N, extractable P and Zn in the former organically manured soils compared to soils treated with mineral fertilizers. The sludge treatment, in contrast to the pig slurry treatment, still showed clearly positive aftereffects on the counts of bacteria, enzyme activities, the mineralization of organic C and N, and on the pH. On the other hand, organic matter content and biological activity were lower in former arable plots than in former grassland plots.

5. CONCLUSIONS

a)   Long-term applications of sewage sludge and pig slurry can increase soil humus content, N content, CEC, heterotrophic soil microorganisms and their activities to a higher degree in a light soil than in a heavy soil. On the other hand, high applications of sewage sludge can decrease autotrophic microorganisms (algae) more pronouncedly in a heavy soil than in a light soil.
b)   Long-term fertilization with sewage sludge can increase soil microbial activities, humus content, N content and pH values not only in the surface soil layer but also in the deeper layers at least down to 1 meter.
c)   Heavy applications of sewage sludge can have lasting positive aftereffects on soil fertility by increasing biological activities and humus content.

ACKNOWLEDGEMENTS

The authors thank S.K. Gupta, C. Gerber, V. Lehmann, M. Roulet, E. Santschi-Fuhrimann, F. Schaer, M. Sievi, A. Schneider and E. Wälti for their helpful contribution in the realization of this work.

REFERENCES

(1)   BECK, T. und SUESS, A.: Der Einfluss von Klärschlamm auf die mikrobielle Tätigkeit im Boden. Z. Pflanzenernähr. Bodenk. 142, 299-309 (1979).

(2)   FURRER, O.J.: Einfluss hoher Gaben an Klärschlamm und Schweinegülle auf Pflanzenertrag und Bodeneigenschaften. Landwirtsch. Forschung Sonderheft 33/I, 249-256 (1977).

(3)   FURRER, O.J., STADELMANN, F.X. und LEHMANN, V.: Die Wirkung steigender Gaben von Klärschlamm und Schweinegülle in Feldversuchen. I. Versuchsprogramm und Auswirkungen auf den Pflanzenertrag. Schweiz. Landw. Fo. 21, 225-237 (1982).

(4)   STADELMANN, F.: Einfluss der Klärschlammdüngung auf die Bodenmikroorganismen und deren Aktivität. In: ALEXANDRE, D. and OTT, H. (Eds.): First European Symposium "Treatment and use of sewage sludge". Cadarache 13-15 February 1979. Commission of the European Communities, pp. 321-329.

(5) STADELMANN, F.X.: Die Wirkung steigender Gaben von Klärschlamm und Schweinegülle in Feldversuchen. II. Auswirkungen auf Population und Aktivität von Bodenmikroorganismen. Schweiz. Landw. Fo. 21, 239-259 (1982).

(6) STADELMANN, F.X. and FURRER, O.J.: Influence of sewage sludge application on organic matter content, micro-organisms and microbial activities of a sandy loam soil. In: CATROUX, G., L'HERMITE, P. and SUESS, E. (Eds.): The influence of sewage sludge application on physical and biological properties of soils. D. Reidel Publishing Company Dordrecht, Boston, London, 1983, pp. 141-166.

(7) STADELMANN, F.X., FURRER, O.J., GUPTA, S.K. und Lischer, P.: Einfluss von Bodeneigenschaften, Bodennutzung und Bodentemperatur auf die N-Mobilisierung von Kulturböden. Z. Pflanzenernaehr. Bodenk. 146, 228-242 (1983).

PHOSPHATE BALANCE IN LONG-TERM SEWAGE SLUDGE AND PIG SLURRY
FERTILIZED FIELD EXPERIMENT

O.J. FURRER and S.K. GUPTA

Swiss Federal Research Station for Agricultural Chemistry and Hygiene of
Environment, CH-3097 Liebefeld-Berne (Switzerland)

Summary

Sewage sludge and pig slurry are organic waste fertillizers, which
contain large quantitities of phosphorus. Their heavy applications can
lead to P-accumualtion in soils. In order to find out the long term
effects of heavy and graded doses of sewage sludge and pig slurry, field
experiments using different soils and different cropping systems were
initiated in 1976. The long term (8 years) effects of heavy application
rates of sewage sludge and pig slurry on the P-uptake, P-balance and
the P-accumulation in soils are reported.
The P-export through plant uptake depends upon soil, application rate
and crop. Specially, in plots receiving heavy application of sewage
sludge about 90% of the totally added P is retained in 0-20 cm surface
layer but a small part is also moved in to deeper layers. Accumulated P
in soils originated from pig slurry is more soluble than that from
sewage sludge.

1. INTRODUCTION

Sewage sludge and pig slurry often lead to overfertilization of soils
with respect to P. The P-content of sewage sludge is high due to an enormous
use of phosphorus in washing powder and to the high efficiency of the
chemical P-elimination in sewage treatment plants. The P-content of pig
slurry is high because pig feed is rich in phosphorus (7-8 g P/kg) due to
abundant use of meat bone meal as a cheap source of protein. Large amounts of
sewage sludge and pig slurry are produced in limited area and and often have
to be disposed off both together in the same area with very limited land
surface, is another reason for P-accumulation in soils.

In field experiments started in 1976 on two different soils (Buren and
Liebefeld) high application rates of sewage sludge and pig slurry were
applied every year in order to find out the long term effects on the soil
properties, the crop characterstics and the leaching loss of nutrients. In
this paper the results, which show the effect of high P- applicaton rates on
the P-content of plants, the P-balance and the P-accumulation in the soil are
presented and discussed.

2. MATERIALS AND METHODS

The field experiments were carried out at two different soils located at :
BUREN: 430 m above sea level ; rainfall 1326 mm, air temperature 8.9 ^{0}C (mean values for 1976-81); gleyed calcareous brown soil, pH 7.3, carbon 2.4%, $CaCO_3$ 25% . It is a silty clay loam (clay 42%, silt 51% and sand 7%).
LIEBEFELD : 565 m above sea level; rainfall 1141 mm, air temperature 8 ^{0}C (mean values for 1976-81); weakly developed parabrown soil, pH 6.3, carbon 2%. It is a sandy loam soil (clay 16%, silt 29%, and sand 55%).

Six different treatments were tested:
0   = control without any fertilization
min = mineral fertilization (150-220 kg N, 50 kg P, 250 kg K/ha).
SS_2 = 2 t/ha organic matter in form of sewage sludge
SS_5 = 5 t/ha organic matter in form of sewage sludge
PS_2 = 2 t/ha organic matter in form of pig slurry
PS_5 = 5 t/ha organic matter in form of pig slurry
Two cropping systems were used:
  Rotation : maize - cereal - clover grass (Plot A, B, C)
  Permanent grassland (Clover grass): (Plot D)

Years	A	B	C	D
1976, 1979, 1982	maize	wheat	grass	permanent grass
1977, 1980, 1983	wheat	grass	maize	permanent grass
1978, 1981, 1984	grass	maize	wheat	permanent grass

Following methods were used to determine soluble phosphorus in the soil samples:
Water-soluble P: 1 g soil, 72 ml water, pH 7
DL -soluble P: 2 g soil, 100 ml(0.04N Ca-lactate+ 0.02N HCl), pH 3.7.
CAL -soluble P: 5 g soil, 100 ml (0.1N Ca-lactate + 0.1N Ca-acetate + 0.3 N
                                  acetic acid solution), pH 4.1.

3. RESULTS

The application rates of phosphorus (Table 1) varied to a large extent among treatments , from no phosphorus in the control treatment to a maximum of annual rate of 440 kg P/ha through the high application of sewage sludge (SS_5). Howevewr, there was insignificant variation in the P-content of the plants due to treatments. The P-export through plant uptake varied to a large extent due to the differences in P-content and in dry matter yield. The highest crop yields were observed in plots which received heavy application of pig slurry. This is due to the fact that it has supplied large amount of plant available nitrogen as compared to sewage sludge.
The P-uptake by plants not only depends on the application rate but also on the type of crop:
silage maize   : 22-45 kg P/ha per year
wheat (grain + straw): 18-35 kg P/ha per year
clover grass   : 37-93 kg P/ha per year
There was large P-uptake by clover grass in the control plot (without any P-fertilization) in both soils even after 8 years of cropping and are comparable to mineral fertilization treatment. The reason for this can be the presence of a large P-reserve at the begining of the experiment in both soil (Table 2). This reserve is surprisingly sufficient even for last 8 years of cropping.

Tab.1: P-content and P-uptake by plants and P-balance of the plots in rotation (maize - wheat - clovergrass). Annual mean values of eight years of experiment. DM = dry matter.

Tretament	0	min	SS2	SS5	PS2	PS5
B U R E N						
P-content (g P/kg DM)						
Maize (tops)	2.44	2.68	2.67	2.69	2.73	2.88
Wheat grain	4.60	4.62	4.70	4.75	4.61	4.74
straw	0.94	1.15	1.17	1.42	1.10	1.39
Clover grass	3.29	3.59	3.64	3.98	3.74	4.13
Yield (t DM/ha)						
Maize (tops)	13.25	15.56	15.51	16.02	15.59	15.79
Wheat grain	3.21	4.11	4.16	4.26	4.23	4.27
straw	3.93	5.95	5.45	5.91	5.51	5.98
Clover grass	11.25	14.92	14.55	16.92	15.19	17.94
P-uptake by plants (kg P/ha)						
Maize (tops)	32.35	41.77	41.38	43.08	42.53	45.52
Wheat grain	14.77	18.98	19.54	20.23	19.49	20.25
straw	3.70	6.83	6.35	8.40	6.04	8.31
total	18.47	25.81	25.89	28.63	25.53	28.56
Clover grass	36.97	53.61	52.95	67.32	56.83	74.12
P-blance (kg P/ha)						
Input (fertilization)	0	50.0	128.8	329.4	72.2	161.8
Export (plant uptake)	29,3	40.4	40.1	46.3	41.6	49.4
Accumulation in soil	-29.3	+9.6	+88.7	+283.1	+30.6	+112.4
L I E B E F E L D						
P-content (g P/kg DM)						
Maize (tops)	1.90	1.99	1.98	2.09	2.00	2.30
Wheat grain	4.65	4.70	4.63	4.80	4.65	4.82
straw	1.04	1.20	1.22	1.54	1.19	1.66
Clover grass	4.06	4.20	4.21	4.40	4.41	4.80
Yield (t DM/ha)						
Maize (tops)	12.04	14.59	14.46	15.25	14.62	15.52
Wheat grain	3.56	4.65	4.43	4.86	4.76	5.01
straw	3.98	6.32	5.01	5.72	5.49	6.53
Clover grass	12.69	15.73	15.32	17.14	17.43	19.45
P-uptake by plants (kg P/ha)						
Maize (tops)	22.88	29.05	28.66	31.94	29.31	35.70
Wheat grain	16.57	21.87	20.52	23.35	22.12	24.15
straw	4.12	7.60	6.13	8.82	6.55	10.86
total	20.69	29.47	26.65	32.17	28.67	35.01
Clover grass	51.52	66.02	64.52	75.37	76.94	93.35
P-blance (kg P/ha)						
Input (fertilization)	0	50.0	179.3	436.9	73.8	174.9
Export (plant uptake)	31.7	41.5	39.9	46.5	45.0	54.7
Accumulation in soil	-31.7	+8.5	+139.4	+390.4	+28.8	120.2

The heavy application rate specially of sewage sludge (SS_5) led to a serious P-accumulation in the soils (Tab. 3). In the undisturbed soil of permanent grass treatment there is very high P-content (over 5000 ppm) in the surface 5 cm as compared to about 1000 ppm in the control treatment. Further there is P-accumulation not only found in the 0-5 cm surface layer but also found even in deeper layers e.g. below 10 cm.

Table 2: Annunal P-application rate, P-uptake by plants and P-accumulation in soil (kg P/ha) of permanent grassland. 8 years mean.

Treatment	0	Min	SS_2	SS_5	PS_2	PS_5
BURREN						
Yield (t DM/ha)	9.48	11.67	11.36	13.14	11.92	13.22
P-content (g P/kg DM)	2.99	3.0	3.52	3.70	3.55	3.78
P-uptake (kg P/ha)	28.3	35.1	40.0	48.6	42.3	50.0
P-fertilization(kg P/ha)	0.0	50.0	134.1	382.5	64.3	161.4
P-accumulation in soil (kg/ha)	-28.3	+14.9	+ 94.1	+333.9	+22.0	+111.4
LIEBEFELD						
Yield (t DM/ha)	11.31	13.52	12.78	13.88	14.7	15.99
P-content (g P/kg DM)	4.61	4.58	4.76	4.63	4.64	4.77
P-uptake (kg P/ha)	52.1	61.9	60.8	64.3	68.3	76.3
P-fertilization(kg P/ha)	0	50.0	185.0	476.6	74.8	189.4
P-accumulation in soil (kg P/ha)	-52.1	+11.9	+124.2	+412.3	+ 6.5	+113.1

The soluble P-fractions were influenced differently than the total P-content. It can be seen from the results presented in Table 3 that the soluble phosphorus contents in plots treated with pig slurry are larger as compared to sewage sludge. This is illustrated by examples of LIEBEFELD and BUREN soil (0-5 cm, Table 3)

Soil	Treatment	P_T	P_W	P_{DL}	P_{CAL}
LIEBEFELD	SS_5	5469	38.2	540	547
	PS_5	1918	94.8	514	624
BUREN	SS_5	5184	20.7	41.3	127
	PS_5	2134	56.4	89.3	125

In the soil BUREN, the soluble P contents are much lower than the LIEBEFELD, which is mainly due to large P-fixation capacity of BUREN soil.

REFERENCES

1. Furrer, O.J., Stadelmann, F.X. and Lehmann, V. (1982). Die Wirkung steigender Gaben von Klärschlamm und Schweinegülle in Feldversuchen 1. Versuchsprogramm und Auswirkungen auf den Pflanzenertrag. Schweiz. Landw. Fo. 21 (3/4) : 225-237.
2. Furrer, O.J. Gupta, S.K. and Stauffer, W. (1984). Sewage sludge as source of phosphorus and conseqences of phosphorus accumulation in soils. In Proceedings of Third International Symposium on "Proceesing and Use of Sewage Sludge" held at Brighton organised by Commission of the European Communities, Brussels. P.LHermite and H. Ott (Editors). D.Reidel Publishing Co. Dordrecht (NL):279-294.

Tab.3: Phosphorus content of the soil under permanent grass after eight (LIEBEFELD) and nine (BUREN) years of experiment. Total (P_T) and phosphorus soluble in water (P_W), lactate (P_{DL}) and acetate-lactate (P_{CAL}).

			mg P_T kg soil	g P_W per t soil	kg P_T	g P_{DL} per t soil	kg P_T	g P_{CAL} per t soil	kg P_T
BUREN	0	0- 5 cm	977	1.88	1.92	0.68	0.70	8.1	8.3
		5-10 cm	865	2.41	2.78	< 0.3	-	< 5	-
		10-20 cm	815	1.97	2.42	< 0.3	-	< 5	-
		20-40 cm	715	1.51	2.11	< 0.3	-	< 5	-
		40-60 cm	559	0.30	0.54	< 0.3	-	< 5	-
	min	0- 5 cm	1457	12.82	8.80	20 .48	14.06	117.2	80.4
		5-10 cm	1125	5.10	4.54	8.87	7.88	49.0	43.6
		10-20 cm	954	3.39	3.55	1.97	2.06	13.9	14.6
		20-40 cm	815	1.26	1.55	4.31	5.30	8.6	10.6
		40-60 cm	630	1.26	2.01	0.3	0.5	5.6	8.9
	SS5	0- 5 cm	5184	20.69	3.99	41.32	7.97	126.9	24.5
		5-10 cm	3465	16.93	4.89	22.56	6.51	102.1	29.5
		10-20 cm	2035	10.02	4.93	11.73	5.76	82.8	40.7
		20-40 cm	1552	2.92	1.88	2.64	1.70	48.5	31.3
		40-60 cm	1061	0.37	0.35	0.45	0.42	18.2	17.1
	PS5	0- 5 cm	2134	56.44	26.44	89.32	41.85	124.8	58.5
		5-10 cm	1684	32.71	19.43	75.93	45.09	113.2	67.2
		10-20 cm	1187	12.89	10.86	10.74	9.05	50.5	42.6
		20-40 cm	999	7.77	7.78	4.93	4.93	32.3	32.4
		40-60 cm	759	1.77	2.33	1.18	1.56	11.1	14.6
LIEBEFELD	0	0- 5 cm	1062	25.0	23.5	97	91	113	107
		5-10 cm	1139	28.5	25.0	119	105	138	121
		10-20 cm	1165	32.4	27.8	160	137	165	142
		20-40 cm	866	29.5	34.1	96	111	127	147
		40-60 cm	696	15.4	22.1	41	59	78	112
	min	0- 5 cm	1330	52.6	39.5	224	168	220	166
		5-10 cm	1160	45.4	39.2	219	189	225	194
		10-20 cm	1124	87.7	78.1	228	204	217	193
		20-40 cm	902	40.6	45.0	158	175	160	177
		40-60 cm	711	22.4	31.5	57	80	100	141
	SS5	0- 5 cm	5469	38.2	7.0	540	99	496	91
		5-10 cm	2694	32.4	12.0	399	148	547	203
		10-20 cm	1717	34.6	20.2	352	205	404	235
		20-40 cm	1061	58.7	55.3	343	323	265	250
		40-60 cm	929	19.1	20.6	90	97	148	160
	PS5	0- 5 cm	1918	94.8	49.4	514	268	524	273
		5-10 cm	1531	77.3	50.5	422	276	375	245
		10-20 cm	1347	73.0	54.2	284	211	365	271
		20-40 cm	1036	59.7	57.6	200	193	250	241
		40-60 cm	704	32.4	46.1	142	202	143	203

CHEMICAL CHARACTERIZATION OF SOIL ORGANIC MATTER IN A FIELD STUDY WITH
SEWAGE SLUDGES AND COMPOSTS

R. Levi-Minzi*, R. Riffaldi**, G. Guidi*** and G. Poggio***
* Institute of Agricultural Chemistry, University of Pisa
** Institute of Agricultural Chemistry, University of Viterbo
*** Institute for Soil Chemistry, C.N.R., Pisa

Summary

Some characteristics of soil organic matter were determined in a field
study established in 1978 on a sandy loam soil planted with corn.
Treatments included a control (C), aerobic (AS) and anaerobic sludge
(ANS), compost of the organic fraction of urban refuse with aerobic
(CAS) and anaerobic sludge (CANS), farmyard manure (FYM) and mineral
fertilizers (MF). All organic materials were applied yearly at a rate
equivalent to 50 tons/ha of manure on an organic carbon basis.
The decomposition of the organic matter added to such a light soil was
fast and only AS was able to increase soil organic-C. AS also
increased total and mineral soil-N. The composition of the humified
fractions of soil organic matter, derived from the determinations of
humic and fulvic acids and gel filtration, was practically the same
for all organic amendments for which the content of fulvic acids was
much higher than that of humic acids. Differences among treatments
were found only for alcohol soluble substances.

1. INTRODUCTION

    The rise and the localisation of most of the population in restricted
areas has forced all developed countries to find methods which keep the
environmental impact of waste disposal as low as possible.
    The application of solid or liquid municipal waste to agricultural
land is generally thought to be the method of disposal which is at the same
time economical and least harmful to the environment. In this way valuable
amounts of nutrients are supplied to plants, and humus forming materials
are added to soil. The latter aspect is of particular importance in Italy
where soils are generally low in organic matter and structural problems are
quite common.
    The aim of this study was to evaluate the differences in some of the
more important characteristics of soil organic matter in an experiment in
which, under natural field conditions, the modifications of soil structure
were previously evaluated (3,6).

2. MATERIALS AND METHODS

    Field trials were established in 1978 at Lamporecchio (Pistoia) on an

Entisol. Some characteristics of the soil at the beginning of the experiment are reported in Table I.

Table I - Some mean characteristics of the soil at the beginning
of the experiment.

Clay (%)	10.0
Silt (%)	14.1
Sand (%)	75.9
Organic-C (%)	0.5
pH (H_2O)	5.8
C.E.C. (meq/100g)	13.4

Plots (500 m^2) were planted with corn and the following treatments were compared: control (C), mineral fertilizers (MF), aerobic (AS) and anaerobic sludge (ANS), compost of the organic fraction of urban refuse with aerobic (CAS) and anaerobic (CANS) sludge, and farmyard manure (FYM). Some analytical mean characteristics of the experimental organic products are presented in Table II.

Table II - Some mean characteristics of experimental organic
materials.

Property	AS	ANS	CAS	CANS	FYM
pH	6.8	6.0	6.6	7.1	6.8
			%		
Water	97.9	90.1	29.0	32.0	75.0
Organic-C *	24.5	38.0	27.8	29.6	31.9
Total -N *	3.5	3.0	2.4	2.5	2.4
Total -P *	3.4	1.1	0.7	0.6	0.5
Total -K *	0.4	0.1	0.9	0.8	1.7

*On a dry matter basis.

Since in this experiment compost and sludge were considered mainly for their content of organic matter, the addition rates were calculated on the carbon basis and were equivalent to 50 metric tons/ha of manure. Organic materials were surface-applied every early spring and ploughed in before the seed-bed preparation. Mineral fertilizers were provided following the usual agricultural practices and were 250, 120 and 120 kg/ha of N, P_2O_5 and

K_2O respectively.

Bulk soil samples, made of five sub-samples, were collected from each plot to a depth of 20 cm in the early October 1983, air dried, sieved out to obtain the fraction less than 2 mm, and used for all determinations. The nitrogen forms were analysed on the fresh soil.

According to the standard methods of soil analysis (2), organic-C was determined by potassium dichromate oxidation, total-N by the Kjeldhal procedure, mineral-N ($N-NH_4+N-NO_3+N-NO_2$) by steam distillation of 2M KCl extracts, hydrolysable-N after an 8 hour reflux with 6N HCl.

Organic matter was extracted by shaking soil samples with 0.5N NaOH (1:10 wt/vol) under nitrogen, overnight at room temperature. After centrifugation, humic acid was separated from fulvic acid by precipitation at pH 1 with H_2SO_4.

The alkali extracts were gel filtrated on Sephadex G-200 by using 0.05N borax as eluant (8). Columns were loaded with 5 ml of extract (about 30 mg of organic-C) and 11.5 ml effluent fractions were collected.

Soil samples were sequentially extracted in Soxhlet extractor with ethyl ether and ethyl alcohol to obtain the following two fractions: i) fats, waxes, oils, and ii) resins (11).

3. RESULTS AND DISCUSSION

In Figures 1 through 8 are presented the data obtained for the main chemical characteristics of the organic matter of the differently amended soils. Histograms followed by the same letter are not significantly different at the P=0.05 probability level by the confidence limits test.

Organic-C - The soil content of organic-C rised only in the plot treated with AS (Fig. 1). The remaining treatments were practically ineffective and the corresponding levels of soil organic-C were not significantly different from each other and also similar to those found in the plots which did not receive any organic amendment.

In a light soil, such as the soil used in this experiment, shortage of oxygen is unlikely to occur and any kind of organic matter is decomposed quickly. For this reason the finding that AS was able to increase the soil organic matter is quite difficult to explain. A possible explanation can be found in the C/N ratios of the different organic materials. The mean C/N ratio was in fact about 7 in AS compared to a value of about 12 for the other organic materials included FYM (5,9). This could mean that the organic matter of AS was already enough humified before the incorporation into the soil and less carbon was available to the oxidation by the soil microorganisms and subsequent loss as carbon dioxide.

Total-N - The patterns of total-N concentrations were similar to those observed for the total-C (Fig. 2). However, in this case the explanation of the major content of total-N found in the plot treated with AS lies in the content of total-N which was the highest in the AS. Moreover, the amount of

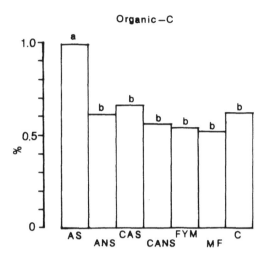

Fig. 1 - Influence of different treatments on soil organic-C.

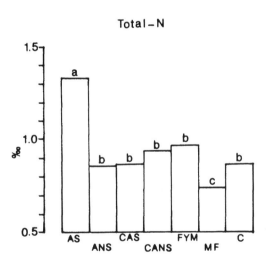

Fig. 2 - Influence of different treatments on soil total-N.

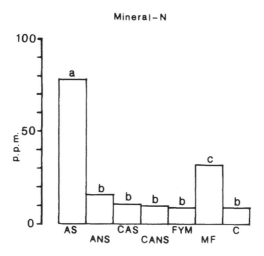

Fig. 3 - Influence of different treatments on soil mineral-N.

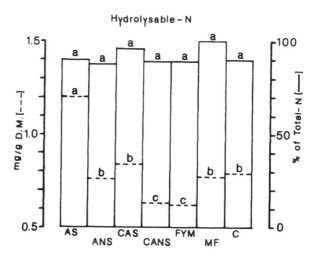

Fig. 4 - Influence of different treatments on soil hydrolysable-N.

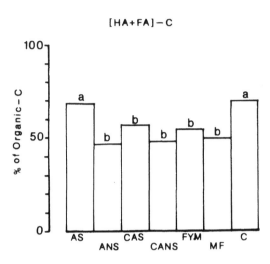

Fig. 5 - Influence of different treatments on soil humified-C.

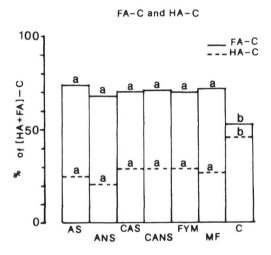

Fig. 6 - Influence of different treatments on the composition of soil
humified-C.

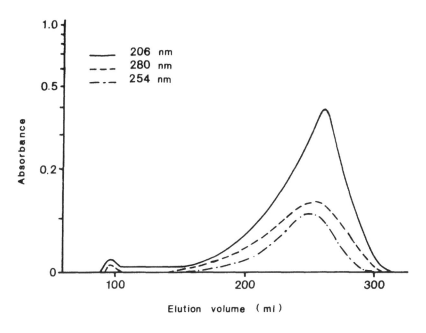

Fig. 7 - Typical elution pattern of alkali soluble soil organic matter.

Ether and Alcohol extracts

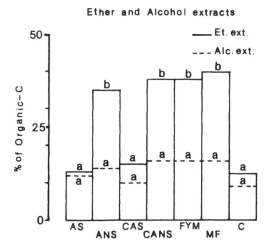

Fig. 8 - Influence of different treatments on the ether and alcohol
extracts from soil.

total-N provided to the soil was further increased because in the AS the mean content of organic-C was lower than in the other materials and, as mentioned before, the treatments were carried out by distributing the same amount of organic-C. The lowest concentration of total-N in the MF plot can be attributed to the fact that the amount of total-N applied with mineral fertilization was less than in the other treatments. Moreover soil microorganisms, the growth of which was not stimulated by the addition of organic matter, were not able to prevent the leaching of the soluble forms of nitrogen through the sandy soil.

Mineral-N - The amount of inorganic forms of nitrogen found in the soil ranged from 1 to 6% of total-N and was practically the same for C, CAS, CANS and FYM (about 10 ppm) (Fig. 3). Higher figures were found for ANS (16 ppm), FM (32 ppm) and expecially for AS (78 ppm). The beneficial effect of nitrogen provided with AS was also confirmed by higher yields of corn observed in the same experimental plot (7).

Hydrolysable-N - According to that observed for the other forms of nitrogen also the hydrolysable-N was higher in the plot treated with AS (Fig. 4). However, if values of hydrolysable-N are expressed as per cent of total-N the highest figure (100%) was found for MF, where the nitrogen was added to the soil only in the mineral form. The determination of hydrolysable-N gives in fact an index of the chemical form of the nitrogen, i.e. the more nitrogen is present in aromatic forms the less nitrogen can be released by hydrolysis.

Humified-C - The organic matter soluble in alkali, (HA+FA)-C, is generally considered as the humified fraction of the soil organic matter. For a better comparison of the results, values of humified-C were expressed in Fig. 5 as per cent of the organic-C. The highest mean percent of humified-C was found in the plot treated with AS. This result agrees with previous findings (Fig.1) and supports the higher resistance of this material to the mineralization in comparison with the other organic materials.

As expected, in the control there is a high percentage of humified material due to the more stable naturally occurring soil organic matter.

In all plots fulvic acid-C represents more than two thirds of the humified-C (Fig. 6); only in C the humus is constituted by fulvic and humic acids roughly in the same proportion. Since to a higher humic/ fulvic acid ratio corresponds a higher humification (10) such a ratio became narrower in the control plot C, in which naturally occurring organic matter had reached a more stable equilibrium.

Gel chromatography of alkali soluble humic substances has been found an useful tool in studying relationships between metals bound to soil organic matter and their availability by plants (8). The distribution of molecular weights of the organic matter extracted from all the experimental plots gave similar chromatographic patterns, an example of which is reported in Fig. 7. Even the use of three different wavelenghts failed to discover differences and therefore also this technique confirms the similarity of the various soil humic substances.

Ether and alcohol extracts - It is well known that ether and alcohol soluble fractions, mainly constituted by waxes, fats, oils and resins, show low decomposition rates in soil (1) and that in sludge treated soils their content tends to increase (4).

In the experiments reported in this paper, patterns of alcohol extracts expressed in percent of organic carbon were not significantly different (Fig. 8).

For what concerns the ether extracts, the C and AS, alone or composted, unexpectedly presented a lower percent of soluble fraction. This may indicate that in some cases, where the humification degree is particularly high, waxes, fats and oils could become incorporated into the soil humus without undergoing major degradative modifications. Such a condensation imparts significant resistance to the extraction by ether (12).

4. CONCLUSIONS

On the basis of the results obtained in the field trials described herein, in which an evaluation was made of the main organic matter characteristics of a soil planted to corn and amended for six years with different organic wastes, the following conclusions can be drawn:

1) The decomposition of the organic matter added to such a light soil is fast and only AS was able to increase soil organic-C. AS also increased total and mineral soil-N.

2) The composition of humified materials in the soil was practically the same for the different organic wastes, with a prevalence of the fulvic acid fraction. However, the trend of humification in the environmental conditions of the experiment is going towards a progressive increase of the humic acid fraction, as appears in the control plot, where the fulvic acid/humic acid ratio becomes about one.

3) The result of the incorporation of different organic materials into the soils originated a humified substance with little differences and showing characteristics comparable to those found for FYM, the typical organic amendment.

5. ACKNOWLEDGEMENT

This research work was supported by CNR, Italy. Special grant I.P.R.A. Sub-project 1. Paper n. 256.

6. REFERENCES

1) Alexander M. Introduction to soil microbiology. Wiley, N.Y. (1961).
2) Black C.A. (Ed.). Methods of soil analysis. Am. Soc. of Agron., Inc., Madison, Wisconsin (U.S.A.), (1965).

3) Guidi G., Pagliai M., and Giachetti M. Modifications of some physical and chemical soil properties following sludge and compost applications. In: Catroux G., L'Hermite P. and Suess E. (Eds.). The influence of sewage sludge application on physical and biological properties of soils. Reidel Publ. Co, Dordrecht, Holland (1983).

4) Hohla G.N., Jones R.L., and Hinesly J.D. The effect of anaerobically digested sewage sludge on organic fractions of Blount silt loam. J. Environ. Qual., 7, 559 (1978).

5) Levi-Minzi R., Riffaldi R., Soldatini G.F., Pini R. Variazioni nel tempo della composizione chimica di fanghi di depurazione. Agrochimica, 25, 168 (1981).

6) Pagliai M., Guidi G., La Marca M., Giachetti M., and Lucamante G. Effects of sewage sludges and composts on soil porosity and aggregation. J. Environ. Qual., 10, 556 (1981).

7) Pardini G., Massantini F. Aspetti agronomici ed energetici della concimazione con "fanghi attivi" a colture inserite in rotazioni diverse. Riv. di Agron., 1, 187 (1983).

8) Petruzzelli G., Guidi G., and Lubrano L. Cadmium occurrence in soil organic matter and its availability to wheat seedlings. Water, Air, Soil Pollut., 8, 393 (1977).

9) Riffaldi R., Levi-Minzi R., and Saviozzi A. Humic fractions of organic wastes. Agric., Ecosys. and Env., 10, 353 (1983).

10) Salfeld J. Chr. and Soechtig H. Composition of the soil organic matter system depending on soil type and land use. Proc. Symp. Soil Organic Matter Studies I.A.E.A., Braunschweig (1976).

11) Stevenson F.J. Gross chemical fractionation of organic matter. In: Black C.A. (Ed.) Methods of Soil Analysis. Am. Soc. of Agron., Inc., Madison, Wisconsin, U.S.A. (1965).

12) Wagner G.H., Muzorewa E.I. Lipids of microbial origin in soil organic matter. Proc. Symp. Soil Organic Matter Studies I.A.E.A., Braunschweig (1976).

DISCUSSION ON PART II

Chairman: Dr J H Voorburg, The Netherlands.

S DE HAAN to I KOSKELA

Are there any sewage treatment plants in Finland where sludges contain
high concentrations of metals?

Answer: There are only a few; in some cases, cadmium and chromium contents
were too high for agricultural use of the sludge. Chromium came from the
leather industry and levels in excess of 5000 ppm in the dry matter have
been measured.

S DE HAAN to I KOSKELA

Referring to the results of experiments conducted in Finland where it was
concluded that larger amounts of sewage sludge could be applied to soils
with a high humus and/or clay contents compared with sandy soils with low
humus contents. Have guidelines been already drawn up in Finland for
practical applications of sludge to land based on these findings?

Answer: The guidelines have not yet been revised to take account of this
fact.

F STADELMANN TO I KOSKELA

In your experiment, as can be seen in Figure 4 you have applied up to 100
tons of sludge dry matter/ha. Is this a normal rate in Finland. In
Switzerland, wecannot accept more than 7.5 tons of sewage sludge dry
matter/ha every 3 years, because of the possible risk of heavy metal
accumulation in the soil.

Answer: Our guidelines in Finland allow us now to use 20 tons dry matter
per hectare every 5th year. It seems that in cases, when soils are rich in
organic and clay material, it is possible to use 40-50 tons sludge dry
matter per hectare once in ten years for example. This would appear to
give the best economical return in terms of fertiliser value and soil
ameliorating effect. The same benefits are not achieved on sandy soils
which require smaller and more frequent amounts of sludge. Heavy metals
present are always taken into account.

J WILLIAMS to P DESTAIN

Is there any possibility that the agronomic benefits from the nitrogen in
slurries applied in the autumn could be assessed more accurately by soil
profile analyses in February; unless this can be used to refine fertilizer
requirements, one is merely guessing at the saving in fertiliser N which
could be achieved in spring.

Answer: Work has commenced on mineral -N in soils after slurry application - but up until now there is little information obtained which could be useful in improving the estimate of slurry N efficiency.

H TUNNEY to P DESTAIN

In the case of slurry applied in the autumn for winter cereals, did you find that the crop was more susceptible to disease?

Answer: We have not observed any greater incidence of disease resulting from slurry dressings in the autumn. Autumn growth of cereals is rather limited in Belgium, except winter barley. The diseases of winter barley observed are principally due to viruses and affect all the experimental plots similarly irrespective of whether they receive slurry or not. For winter wheat there is no great disease problem as the crop is usually sown in late autumn normally after sugar beet harvest.

O FURRER (Comment)

Numerous experiments in Switzerland showed that the value of slurry as a nitrogen source is high due to the high soluble N content and the good availability. We have to distinguish between availability and efficiency. The availability depends mainly on the portion of mineral N ($NH+_4$) in the slurry, which is just as available as N from mineral fertilizer. In contrast, the organic N is only about 20% available in the first year after application.

The efficiency depends on the application rate (the same is true for mineral fertilizers) and is influenced mainly by the extent of the losses by NH_3 volatilisation, which are high at high pH, dry and hot weather and high NH_4 - contents (biogas - slurry), by denitrification and by leaching.

In order to have good efficiency, it is important to apply the right amount at the right time, to an appropriate crop in a favourable manner.

J HALL to K SMITH (Comment)

It was found in trials in the West of Scotland where annual rainfall is in excess of 1000 mm that thin sludges of 2-4% dry solids were more effective during the summer months and that the efficiency of thick sludges were only fully realised during the wet seasons.

H TUNNEY to K SMITH

An efficiency of 30% for slurry N seems rather low but perhaps correct for many situations. Why do you think this should be so?

Answer: The efficiency of the nitrogen in cattle slurry obtained in UK trials, has been extremely variable tending towards 30%, as illustrated in our latest (MAFF) series of trials. The figure has been recently reduced from 50% to 30% in our advisory literature. In individual cases, much higher efficiencies are obtained and can be explained by one, or sometimes more factors eg soil type, cumulative rainfall following application and temperatures.

Higher slurry N efficiencies are usually recorded in pig slurries but I feel that the 30% efficiency figure is about right in many cases, bearing in mind the importance of farmers applying adequate N for maximum crop response.

S DE HAAN to K SMITH

The better availability of N in pig slurry compared with cattle slurry could be due to the fact that pig feed is much richer in protein and consequently the organic matter in the faeces is more biodegradable – would you agree?

Answer: I would agree with your suggestion. Feed undergoes a more thorough digestion in the digestive tract of ruminants than other animals such as pigs. Therefore the residual organic material would be expected to be less biodegradable. I think there was a parallel situation described in the paper by Mr Hall where it was pointed out that a higher proportion of the organic N is available in undigested sludges compared to digested sludges.

L BARIDEAU to F STADELMANN

Earthworms probably played an imporant role in the admixing of the top layers in the profile. Did you make any observations on the activity of earthworms in this trial and were there any differences to be found between the treatments?

Answer: Up until now we have not examined the influence of sewage sludge and pig slurry on the numbers and activity of earthworms in the soil. But semiquantitative observations during 6 years (enumeration of the earthworms in the taken soil samples) lead to the conclusion that, in sludge and slurry treated soils, more earthworms were present than in untreated soils. This is due to the direct addition of organic matter and/or indirectly due to higher root residues. The increased contents of nitrogen, phosphorous and organic matter in the subsurface soil (30-100 cm) after sewage sludge application could not be explained simply by physico-chemical translocation of the nutrients and organic matter to deeper soil layers. Active transport by earthworms is a much more likely explanation.

G GUIDI to F STADELMANN

Why should the increase in organic C be greater in a light than in a heavy soil? Decomposition is normally faster in light textured sandy soils because of better aeration?

Answer: Organic matter applied in form of sewage sludge or pig slurry was decomposed quicker in a light than in a heavy soil, which can be seen in Table 1 (C – mineralization). Nevertheless, a yearly application of organic manures and wastes during seven years increased more markedly the organic C in a light than in a heavy soil because

- the initial organic C content of the light soil was on a lower level than in the heavy soil

- without manures (treatment 0) the organic C content decreased more rapidly in the light than in the heavy soil

- sewage sludge and pig slurry applications increased grass yields more in the light than in the heavy soil. This meant, that the influence on the root production as a source of organic C was more pronounced in the light than in the heavy soil.

S COPPOLA to F STADELMANN

You did not investigate the long term effects on nitrification and denitrification and I wonder whether you have any comment to make on this aspect.

Answer: Many years after the last sewage sludge applications, it was possible to find higher counts of NH_4^+ - oxidizers, NO_2^- - oxidizers and denifrifiers in the soil. But it is improbable that some years after the last manure or sludge applications, significantly higher nitrification and denitrification rates can be observed. Because after some years, the easily decomposable organic substances of sewage sludge and slurry, which are energy sources for the denitrifiers, are no longer present in sufficient quantity. However, it is certain that directly after the sludge and slurry applications, appreciable nitrogen losses due to increased nitrification and denitrification rates can be expected.

A SUSS to O FURRER

What is the significance of the high soluble phosphorus levels achieved in soils after sludge and slurry application?

Answer: The high soil P contents was not a risk for plants; in sludge also high amounts of trace elements (Zn!) are applied which reduce P availability. But high P content in the surface soil is a danger for water pollution by erosion (entrophication of lakes). Too high P accumulation in soil should be avoided for economic and ecological reasons. Continuous P accumulation in some soils over a long period will result in increased leaching losses of P.

A SUSS to O FURRER

Which methods were used for the measurement of nitrate in the ground waters and what were the rates of nitrate leaching in these experiments?

Answer: The water samples were taken from 1 metre depth in the soil by the porous cup method. The groundwater was not sampled, as this was only to be found below 20 m depth. Nitrate leaching was very low in the permanent grassland, where the measurements were made.

B POMMEL to O FURRER

Is it possible to explain the differences obtained between sewage sludge and pig slurry?

Answer: A possible explanation was that the iron bound sludge P was less soluble than P bound with easily decomposable organic matter of pig slurry.

J HALL to O FURRER

Is there any reason for the lower solubility of sludge P compared to pig slurry?

Answer: Most of the total P present in sewage sludge used for this experiment is Fe-bound P of low solubility. The sludge used was taken from a waste water treatment plant having the facility of P elimination by addition of the salts and that, in the pig slurry, appreciable amounts of P are bound to organic material, which is easily decomposable.

J VORBURG to G GUIDI

Would you please explain in more detail the composition of the composts which you used in your experiments.

Answer: The CAS was made by composting together the organic fraction of urban refuse and aerobic sludge (60-40% W/W) and CANS similarly but using anaerobic sludge (80-20% W/W).

S de HAAN to G GUIDI

Can you explain the reason for such a small increase in the organic C content of soil when 25 t/ha of carbon was applied in sludge to a soil with an initial amount of organic carbon not more than 15 t/ha. A doubling of the C content should be expected as was the case only with aerobically treated sludge?

Answer: Firstly I would like to explain that the plough layer for the maize crop in Italy is about 40 cm deep and consequently the content of soil organic-C is about 22 t/ha. This figure is therefore similar and not much lower than that of the total organic-C added to soil since 1978. This may partly explain the generalised lack of increase of soil organic matter because the amount of organic carbon added with organic amendments is lower in percentage and is incorporated in a deeper layer of soil. Moreover, in addition to the observations already made by Dr Pagliai on the decomposition of soil organic matter in Italy, I would like to remind you that, in cultivated soil, an annual loss of 10 t/ha of organic matter is not unusual.

K LARSEN to L C de la L CREMER

Did the rates used, beginning with 100t cattle slurry/ha/yr going up to 300t/ha/yr give any problems with stock health?

Answer: The quantities of slurry used in the experiments covered rates from efficient manuring to dumping of slurries. There can be a relationship between stock health and too heavy or frequent use of animal waste, for example, by too high K or NO_3-N contents in the fodder or high Cu-concentrations (especially in slurries from fattening pigs) The effects of an excess of K can be controlled by magnesium supplementation.

K SMITH to L C de la L CREMER

Your results illustrated a positive "rest effect" from slurries with
mineral fertiliser applied in combination with the slurry. They also
showed some negative rest effects. From this can we therefore infer that
the overall rest effect is negligible? Our own MAFF experiments in the UK
have tended to indicate a purely additive effect of slurry -N and
fertiliser -N.

Answer: The "rest effect" which is the difference in maximum yield that
can be obtained between the fertilizer system and the fertilizer + manure
system is an interesting one and not clearly understood. We estimate that
some 15 different factors are involved, the intensity of which can vary
from year to year. It is not a simple matter of addition of effects of
fertilizer N and organic matter N. It happens on a larger scale on arable
land, but is also found on grassland. It is not just a first year effect
but occurs also as a residual effect on land treated regularly with organic
manures. The yield differences can be very variable. Negative effects may
occur following failures by ploughing in the organic matter for example,
under wet conditions (H_2S formation, insufficient soluble N for straw).

B. CHANGES ON STORAGE AND MINERALISATION STUDIES

IN SOILS AFTER TREATMENT

The effect of storage on the utilisation of sewage
sludge

Mineralisation of organic matter in soil treated
with sewage sludge stabilized by different methods

Sludge origins and nitrogen efficiency

Discussion

THE EFFECT OF STORAGE ON THE UTILISATION OF SEWAGE SLUDGE
E. VIGERUST
Department of Soil Fertility and Management
Agricultural University of Norway

Summary

During storage different characteristics of sludge can be changed.
Samples of stored sludge for analysis are taken at different times
and from different depths. The percent of dry matter and ash increase
during storage, especially in the surface layer. The water loss brings
about a certain shrinkage of the volume. Decomposition decreased the
C/N-ratio. The percent of total nitrogen was not much changed by storage. A certain loss of N is, however, registered, and this represents
the part of the nitrogen which most easily can be utilized by the
plants.
Experiments showed that composting, more than storage, reduced the
effect of sludge compared to fresh sludge. Storage of raw sludge some
months can prevent phytotoxic effects.

1.   INTRODUCTION

Under the climatic conditions in the nordic countries, dewatering of
sludge is necessary. Compared to Central Europe the spring is short and
the spring work has to start as soon as the soil is dry enough. The most
actual time for sludge spreading is, therefore, in the autumn.
    Distribution of sludge is regarded as difficult, and sludge storage
is an important part of this problem. Some actual alternatives are shown
in this figure:

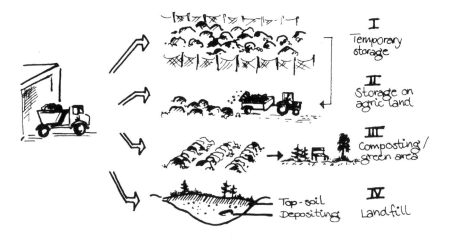

I Temporary storage

II Storage on agric. land

III Composting / green area

IV Landfill

Top-soil Depositing

To provide a safe storage of sludge minimizing the risks of water pollution and hygienic problems, a few controlled central places are regarded as most suitable for this purpose; places for temporary sludge storage (alt. I).

If the sludge has to be loaded once more into trucks for transportation to agricultural land, the expenses easily will be too high for a common agricultural utilization.

More and more common is the transportation of sludge directly to the farms, where it is stored until it can be spread (alt. II). This storage has to be accepted by health authorities.

Another alternative is sludge composting (III) for use on green area. A certain part of the sludge is brought to landfills (alt. IV). It can be deposited in the landfill but sludge can also be stored and later used as part of the top soil when the landfill is finished.

2. CHANGES IN SLUDGE QUALITY DURING STORAGE

During storage the sludge gradually changes. How fast and thorough these changes take place depends on the storage time, climate and sludge properties, especially percent dry matter (DM). The drying up of the surface layer is mainly due to evaporation. Degradation of organic matter increases and consequently increases the aeration, as indicated in this figure:

Fresh sludge   Stored some months   Stored 1-3 years

2.1 Dry matter content

Due to pressure a part of the water will be pressed out from stored sludge. A higher dry matter content gives a greater possibility for microbiological activity and increased evaporation.

Sludge samples taken, at different times and depths, from heaps of 6-10 m^3 were analysed. The analyses results of freshly heaped sludge (April, A) and after 6 months storage (September, B) are shown in table 1.

All our investigations show that DM-content, after short time, increases to 26-29 pst. in the deeper part of stored sludge. This waterloss (presswater) is, therefore, high from a little dewatered sludge. An increase in DM-content from 20 to 26 pst. represents a waterloss of about 230 1 per m^3. Analysis of leaching water indicates the following losses of NH_4-N, P and organic matter through presswater from a 20 pst. DM sludge:

	NH_4-N g/m^3	P g/m^3	Org. matter g/m^3 as CoD
Raw sludge	310	0.006	2780
Anaerobic digested	190	0.008	210

Although precipitation also contributes to the leaching losses, this can be more pronounced in well "open" and decomposed sludge. Fresh sludge

Table 1. Chemical composition of sludge samples taken at different depths. A: Fresh, B: Stored from April – September

Depth cm	DM %	Org.m. %	pH	Cond. µS	tot-N %	tot C %	C/N	"tot-N"* %	NH_4-N mg/100 g DM	NO_3-N mg/100 g DM	P-Al mg/100 g
A (average)	23.3	58	7.4	2274	2.02			2.02	555	0	
B											
0-10	35.6	46	5.8	883	1.99	21.8	10.6	1.65	37	63	239
15-25	29.7	50	6.1	1080	1.91	25.9	13.5	1.70	269	40	272
30-40	27.5	53	6.9	1475	2.11	27.7	13.7	1.98	396	18	295
50-60	26.3	56	7.3	1728	2.07	29.1	15.0	1.97	589	2	306
Parallel obs.	21	21	21	21	16	6	6	10	21	21	6
F-value	15.0***	18.0***	12.2***	7.4***	1.9	8.0**	14.0***	5.6**	7.23***	10.3***	15.1***

* "tot-N" = tot N calculated as percent of the original sludge DM.

Table 2. Chemical composition of stored sludge without and with vegetation, average of samples from 4 depths

	DM %	Org.m. %	pH	Cond. µS	tot-N %	tot C %	C/N	NH_4-N mg/100 g DM	NO_3-N mg/100 g DM
Without veg.	27.8	51	6.6	1396	1.94	24.7	13.8	289	19
With veg.	33.0	51	6.4	1620	2.11	27.3	12.6	332	47
Parallel obs.	27	27	19	27	23	8	8	27	27
F-value	9.1***	2.1	1.3	1.3	2.1	1.61	2.0	1.57	6.2**

is usually less permeable to water and is, therefore, less susceptible to leaching.

2.2   Shrinkage

Loss of the sludge-bound water results in the shrinkage of the sludge, depending among other things, on the dry matter content. The following figure shows the change in volume upon drying of sludge samples in laboratory:

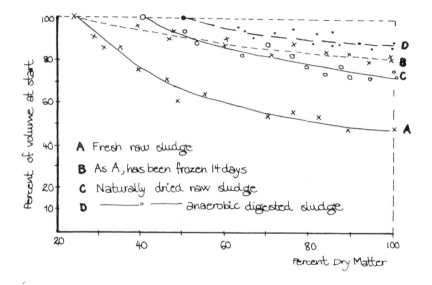

The volume at the start was 100 pst. After a period of drying there was a marked decrease in volume for the little dewatered sludge (can shrink much). Rapid drying of the top layer of moist sludge can give a very hard crust which can cause problems for spreading. Frost in the sludge down to 30-40 cm will later give a loose and well aerated sludge.

2.3   Organic matter

During storage the organic matter decomposes especially in the upper layer (table 1). Different types of sludge behave differently in this regard. Sludge dewatered to more than 27-30 pst. DM can be decomposed at high temperatures in the upper 30-50 cm. Composting by turning, after a period of drying, can be a practical method. Decomposition of sludge changes the physical properties e.g. permeability of water.

2.4   Nitrogen

The total N-content is normally not much changed during storage (table 1). As a result of degradation the C/N ratio decreases. The loss of organic matter results in a corresponding decrease in mass. The total-N analyses are, therefore, also calculated in pst. of the original dry matter content, "tot-N" in table 1. On average the N-loss during the storage period is calculated to 91 pst. of tot-N content in spring. This represents approximately 2 kg N per ton DM. This is, however, just the part which most easily can be utilized by the plants.

For fresh sludge different formulae are proposed in order to calculate the nitrogen effect of sludge expressed as pst. N of the DM-content, viz.:

$$0.9 \times NH_4\text{-}N \text{ pst.} + 0.25 \times \text{org. N pst.} \tag{1}$$

Such formulae can scarcely give an adequate measure of the nitrogen effect for stored sludge, especially because sampling and analyses in the storage period is practically difficult.

The content of NH_4-N is much reduced in the upper layers after storage for some months. Nitrification takes place in well decomposed material, and gradually reaches greater dimensions.

The losses of nitrogen as NH_3 depend on how "open" the sludge is, the degree of decomposition and pH. In a similar investigation with 3 lime-treated sludges, dewatered to about 30 pst. DM, the nitrogen loss during storage from April to September was calculated to 42 percent of the original N-content (2).

2.5 Phosphorus

In Norway plant available P in soil is determined by extracting with ammonium-lactate (AL-method). Drying resulted in the decrease of the P-AL content, probably because of oxydation of iron (3).

2.6 Establishment of vegetation on stored sludge

Sludge which is stored or deposited might lead to water pollution. By establishment of vegetation on the sludge evaporation can be increased and the amount of leaching water considerably reduced. A change in some chemical properties will also take place (as shown in table 2). Experience has shown that it is possible to grow rye grass or fodder rape on pure raw sludge. This can be used on landfills or as a pretreatment for sludge composting.

3. EFFECT OF SLUDGE STORAGE ON GROWING CONDITIONS

In a pot experiment the effect of fresh sludge, composted sludge, surface layer and central layer of sludge stored one year were compared. Amounts equivalent to 80 tons DM/hectare were applied. The total yield of rye-grass, 5 cuttings, is shown in table 3.

Fresh sludge had the best effect, while the effect of compost was rather little. At the first harvest the surface layer of stored sludge and compost gave the highest yields. The less decomposted sludge had in this period probably some phytotoxic effects to plants, as described by ZUCCONI et al. (4). As a whole the central layer of stored sludge had better nitrogen effect than the sludge from the surface.

On the other hand, experience shows that it is easier to spread stored sludge with machines than the fresh types.

In another pot experiment grass was grown in pure sludge with different degrees of decomposition. Chemical analysis of the sludges and yield response are expressed in table 4.

Table 3. Pot experiment where different sludges from the same purification
plant were applied, 80 tons DM/hectare. Yield of rye-grass g per
pot

| | Control | Fresh sludge | Stored sludge | | Sludge compost |
			Surface	Center	
N_o	28.1	73.7	52.7	56.7	42.9
N_1	50.7	87.7	67.1	66.4	61.5

N_o: Without N-fertilizer.
N_1: 200 kg N/hectare (40 kg x 5 applications).

From an experiment where 40 tons sludge/hectare were applied.
From left control, fresh sludge, surface layer of stored sludge
and central layer of stored sludge.

Degradation of organic matter reduces some of the harmful effects of
sludge such as high content of NH_4-N, phytotoxic effects from intermediate
humus compounds and a dense structure. Even composted pure sludge will have
sufficient nitrogen content for maximum growth.

In our pot experiments, we regularly observed plant injury, the first
1-2 months after application of raw sludge in amounts of 20-60 tons DM per
hectare. Some plant species e.g. rye-grass and fodder-rape tolerate little
decomposed sludge better than other species, e.g. cereals, which are more
sensitive.

The same harmful effects are in a less degree observed in field experiments. The inhibiting effect in the field seems to be less and during
the whole growing season this can be overshadowed by the positive effects
from sludge.

During storage for some months, the sludge can be enough decomposed
to prevent visible plant injury under field conditions.

A field experiment is carried out with increased applications of different sludge types, viz. (see fig. 4):

A Fresh anaerobic digested sludge
B Fresh lime-treated sludge (not digested), dewatered to 40 pst. DM
C Sludge compost (ripe), same origin as B.

Table 4. Chemical composition of sludge and yield of grass grown in pure
sludge with different degrees of decomposition

Storage time months	Depth cm	DM %	Org. matter %	NH₄-N	NO₃-N	Yield g/pot Harvest 1st	2nd	3rd
				mg/100 g DM				
A 6	0-10	35	42	13	56	2.9	13.6	15.1
	20-25	33	43	448	6	1.8	14.0	17.7
	40-50	30	51	83	0.4	0	0	6.2
24	0-10	42	33	8	100	3.7	12.5	16.3
	40-50	37	45	175	52	4.4	17.4	18.4
Composted		45	34	152	0.2	3.4	16.5	20.6
B 4	0-10	29	45	442	0.2	2.8	17.3	9.9
	30-40	25	46	777	0.1	0	0.6	6.4
9	0-10	32	44	15	97	5.6	22.7	19.5
	30-40	29	47	880	0.3	0.2	16.4	18.5
20	0-10	41	41	13	53	3.7	18.4	18.9
	30-40	49	36	108	68	3.8	23.3	23.3

Sludge: A) Raw sludge   F-value: 16.2   18.8   11.0
        B) Anaerobic digested Lsd: 1.2   5.8   6.0

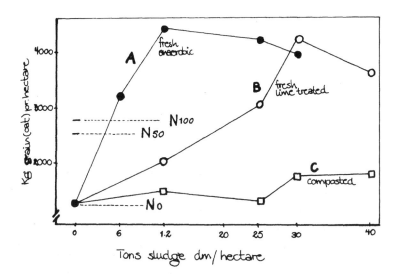

Fig. 4. Yield of oat at increasing application of different sludges

The response measured as yield of oat (grain) is shown in fig. 4.
The yield upon application of NO, N50, and N100, (0, 50 and 100 kg N per
hectare) without sludge is shown by the broken lines.
    Composted sludge has a loose and very good structure for green area.
The value for agricultural purpose is, however, less. Application of lime
usually catalyses the loss of NH_3. Well dewatered lime treated sludge,
after a short time looses much nitrogen amplified by the decomposition of
organic matter during storage, which for a well dewatered sludge also takes
place during the winter season.

4. DISCUSSION

    It is difficult to establish a sampling system for stored sludge for
chemical analysis. The analysis results may be difficult to interpret and
be used in advising farmers on the need for additional fertilizer application. Analysis of fresh sludge, the length of storage time and properties
of the sludge are important criteria for the stipulation of the N-fertilizer effect. From a practical point of view guidance ought to be given on
the basis of volume, because it is difficult for the farmers to ?certain
    the dry matter weight of the sludge. During storage a greater reduction
of volume than of dry weight and N-amount, occurs. The N effect of the
sludge is, however, reduced to a greater extent than what the N-loss during
storage indicates. These changes in volume and N content are on the other
hand, dependent upon various properties of the sludge, such as DM content,
structure, etc. More detailed studies on this field are started.
    In Norway large amounts of sludge - not biologically digested - are
produced. To safeguard against injurious effects of e.g. phytotoxic nature,
well decomposed sludge is recommended. However, higher yields can be obtained by using less decomposed sludge. This type is more effective in pre-
serving the N-content and there are clear indications that the positive
effects of sludge on soil physical conditions are greater for fresh organic
matters than the more decomposed variant. Use of this type requires, however, a sound knowledge on soil type, choice of plant type, and amounts to
be applied.
    It is generally experienced in Norway that decomposition is essential
to how the plants react. Moreover, a considerable reduction of vital cys-
tes of eelworms and other infectious diseases can be attained.
    Sludge storage is a sensitive area and can bring about a lot of con-
flicts between neighbours as the figure below tries to show:

5. LITERATURE

1. FURRER, O.J. und R. BOLLIGER. 1979 Wirksamkeit des Stickstoffs von verschiedenen Klärschlämmen im vergleich zu Ammonitrat. Symposium "Treatment and Use of Sewage Sludge" Cadarache 1979.

2. VIGERUST, E. og J.G. WENG. 1983. Kalkholdig kloakkslam - Metoder, virkning og bruk. Rapp. B 4/83 Inst. f. jordkultur, NLH, 31 s.

3. VIGERUST, E. 1983. Gjødseleffekt av fosfor i kloakkslam. Rapp. B 2/83 Inst. f. jordkultur, NLH, 24 s.

4. ZUCCONI, F., A. PERA, M. FORTE and M. DE BERTOLDI. 1981. Evaluating Toxicity of Immature Compost. BioCycle, March/April. p. 54-57.

MINERALISATION OF ORGANIC MATTER IN SOIL TREATED WITH
SEWAGE SLUDGE STABILIZED BY DIFFERENT METHODS

S.DUMONTET, E.PARENTE and S.COPPOLA
Istituto di Microbiologia agraria e Stazione di Microbiologia industriale
Università di Napoli - I 80055 Portici, Italia

Summary

Organic matter mineralisation has been investigated in soil treated
with the same sludge, stabilized by different processes: liquid
aerobically digested; dewatered and composted in mixture with wood
chips; dewatered and composted with inert bulking agents.
The same amount of organic matter has been applicated to soil in
the form of differently stabilized sludge. Carbon and Nitrogen
mineralisation has been monitored during incubation as well as
field experiments. Variations of the microbial groups responsible
for mineralisation processes have been evaluated. At the end of
trial, soil biomass and FDA-hydrolytic activity have been
quantified. The study allowed to evidentiate differences among
mineralisation rates of the various materials.

1. INTRODUCTION

Techniques used in sewage sludge stabilisation provide different
final products, responsible for different effects after their application
to agricultural land. Liquid digestion, lime treatment, composting a.s.o.
produce sludges characterized by different physical, chemical and biologic
al properties. Agronomic performance, application rates and environmental
effects can consequently vary.
A wide and detailed experience is available as fas as practical,
technical and scientific aspects of agricultural utilization of liquid
sludges are concerned. Much more rare studies regard the effect of soil
treatment with sludge stabilized in the solid phase. Intensity and lenght
of the effect of sludge application widely depend on the quality of organic
matter, therefore on the method of stabilisation.
In this paper results obtained comparing the behaviour in the soil of
the same sewage sludge stabilized by different techniques are reported.
Mineralisation rates and influences upon soil microflora have been parti
cularly considered through field and laboratory experiments.

2. MATERIALS AND METHODS

Sewage sludge from the waste water treatment plant of Torre del Greco

-Villa Inglese (25,000 inhabitants, Province of Naples) has been stabilized by three different methods: 1) liquid aerobic digestion; 2) composting in mixture with wood chips; 3) aerobic stabilization in the solid phase in mixture with synthetic and inert bulking agents. Characteristics of sludge and methods of stabilisation have been previously reported (4).

Experiments have been carried on a volcanic sandy-loam soil of Naples area (pH (KCl)=7.1; organic Carbon, 30.38 mg.g^{-1}; total Nitrogen, 3,167.3 mg.Kg^{-1}). Treatments provided equivalent amounts of organic matter at different rates, in the form od the various sludges. Liquid sludge applica tion was the first of four treatments. Amounts of dry and organic matter, as well as of nutrients applicated are reported in table 1.

Table 1. Application rate of organic and inorganic substances in the form of sludges stabilized by different methods

FIELD TRIALS			
	liquid digested	composted in mixture with	
		wood chips	inert agents
Organic matter (t/ha)	9.0	36.0	36.0
Dry matter "	15.7	47.0	72.6
Organic Carbon "	4.6	15.4	14.89
Total Nitrogen (Kg/ha)	361	630	2,620
Inorganic Nitrogen "	65	55	44
Organic Nitrogen "	298	605	2,576

LABORATORY TRIALS				
	liquid digested	composted in mixture with		
		wood chips		inert agents
		Rate I	Rate II	Rate I	Rate II
Organic matter (g/Kg of soil)	2.3	2.3	9.2	2.3	9.2
Dry matter "	4.03	3.08	12.3	4.65	18.61
Organic Carbon "	1.17	0.98	3.93	0.95	3.81
Total Nitrogen (mg/Kg of soil)	92.69	41.5	166.0	167.0	672.0
Inorganic Nitrogen "	16.19	1.68	6.71	2.71	11.16
Organic Nitrogen "	76.49	39.82	159.29	164.21	660.84

Incubation studies have been carried out at 30°C and soil water content was brought to 60% of field capacity. Organic Carbon and total Nitrogen have been determined according to routinary methods (2). NH_4^+ was detected in 1 M KCl extracts according to a modification of Berthelot (10) method; NO_3 by brucine method (8). A modification of Jenkinson's technique for the evaluation of soil microbial biomass has been utilized (3). Hydrolytic activity on fluorescein diacetate in soil has been investigated according to Schnürer and Rosswall (7). The most probable number (MPN) of ammonifier microorganisms in control and treated soils has been quantified according to Pochon and Tardieux (6), employing Casaminoacids (Difco Lab., USA) instead of asparagine in the medium. NH_4^+-oxidizers were studied by the technique of Soriano and Walker (9); NO_2-oxidizers according to Alexander and Clark (1). Soil respiration was monitorized by entrapping CO_2 in 0.25 N NaOH, precipitating the carbonate with 20% $BaCl_2$ and titrating the excess base with 0.1 N HCl in the presence of thymolphtalein as an indicator.

3. RESULTS AND DISCUSSION

   Differences among mineralisation rates of organic Carbon from the various materials utilized are shown in figure 1, in which: O=control soil; ●=liquid sludge; △=sludge stabilized in mixture with inert bulking agents rate I; ▲=the same, rate II; □=sludge composted in mixture with wood chips, rate I; ■=the same, rate II.

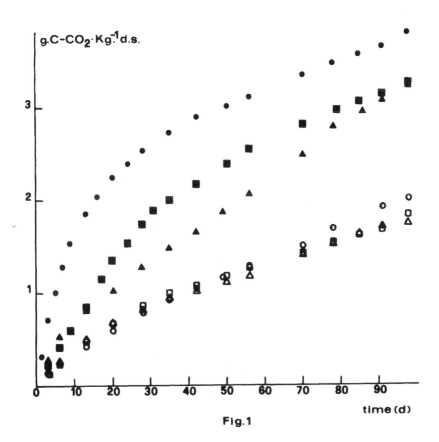

Fig.1

The figure regards the incubation trials. In these experiments Carbon mine_
ralisation is far resulted faster for liquid sludge, although lower amounts
of organic matter have been applicated by this treatment.The process does
not show significant differences in the control soil as well as in the soil
treated with lower application rates of composted sludges. On the contrary
it shows middle courses with the two highest compost applications, without
significant difference between the sludge stabilized in mixture with inert
bulking agents and the sludge composted with wood chips.The results of
these analyses and their statistical evaluations are summarized in table 2.

Table 2. Mineralisation of organic Carbon in soil treated with sludge
stabilized by different methods (incubation experiments)

| Treatments | g.Kg^{-1} of dry soil | | | | |
	Carbon applicated	Total C in soil	C-CO_2 released[*] after 98 days	% mineralized of C applicated[**]	total
1) Control	–	30.380	1.966	–	6.47
2) Liquid sludge	1.17	31.550	3.826	158.97	12.13
3) Sludge composted with wood chips Rate I	0.98	31.362	1.849	n.s.	5.90
4) Sludge composted with wood chips Rate II	3.93	34.310	3.278	33.38	9.55
5) Sludge stabilized with inerts Rate I	0.95	31.331	1.744	n.s.	5.57
6) Sludge stabilized with inerts Rate II	3.81	34.190	3.302	35.06	9.66

[**] = Corrected for respiration of control soil.
[*] = F significant at 1% level

Newman and Keuls' Test (P=0.05)

| 5 | 3 | 1 | 6 | 4 | 2 |

They put into evidence that an amount of organic Carbon higher than the
applicated one has been mineralized after the treatment with liquid sludge
(positive "priming effect"). Moreover data analysis confirms that the bio_
logical activity of soil treated with the inferior rates of composts does
not differ from control in the period considered; whereas the highest rates
give rise to mineralisation of about 35% of the organic Carbon applicated
in 98 days, corrected for control respiration.

Respiration data of soil samples treated with the various sludges have
been utilized attempting to achieve a model of the phenomenon. A relation
proposed by Chaussod and Nicolardot (3) to support the evaluation of micro_

bial biomass according to Jenkinson's method has been chosen. This relation
is expressed by the equation

$$y = c \, t + a \, (\, 1 - e^{-K_a \, t} \,)$$

where:
y = CO_2 released at the time t, as Carbon
c = rate constant of zero-order
a = organic Carbon mineralizable according to the first-order kinetics
K = rate constant of first-order.

It assumes that organic Carbon in soil samples includes a part mineraliza
ble according to a zero-order kinetics (part I) and a part mineralizable
according to a first-order kinetics (part II). Estimated values of c, a
and K_a (table 3) show that soil samples treated with sewage sludge stabili
zed in mixture with inert bulking agents don't differ between themselves
and in comparison with control, as regards the part II of organic Carbon.

Table 3. Estimated values of the coefficients of the respiration equation
for soils treated with sludges stabilized by different methods.

Treatments	c (mg C.d^{-1})	a (mg C.Kg^{-1}dry soil)	K_a (d^{-1})
1) Control	15.30±1.02	470.85±91.84	0.050±0.013
2) Liquid sludge	16.86±0.47	2,169.94±32.58	0.110±0.003
3) Sludge composted with wood chips Rate I	12.01±0.71	622.63±59.35	0.061±0.009
4) Sludge composted with wood chips Rate II	8.19±2.53	2,613.94±295.84	0.020±0.003
5) Sludge stabilized with inerts Rate I	13.46±0.38	461.54±28.28	0.970±0.013
6) Sludge stabilized with inerts Rate II	28.95±0.35	462.29±22.73	0.210±0.033

Their rate constats K_a are on the contrary strongly affected by application
rate. A remarkable dose-response is shown by soils treated with sludge
composted in mixture with wood chips. Soil samples treated with liquid
sludge present a very high value of organic Carbon mineralizable according
to a first-order kinetics characterized by the highest rate constant.
Mineralisation rate of organic Carbon belonging to the part I (zero-order
kinetics) is considerably affected by the highest doses of composted
sludges only: the sludge composted with inerts shows the highest value of
the constant c; the sludge composted with wood chips, the lowest one.

The analysis of organic Carbon contents of soil treated in field
experiments doesn't significantly elucidate the process, like data of table

4 show.

Table 4. Organic Carbon contents (mg.g^{-1} dry soil) of soils treated with sewage sludges stabilized by different methods

Time (months)	Treatments			
	Control (A)	Liquid sludge (B)	Sludge composted with wood chips (C)	Sludge stabilized with inerts (D)
0	28.59$^{\pm}$0.33	30.13$^{\pm}$0.61	39.19$^{\pm}$2.53	33.02$^{\pm}$1.68
2	29.64$^{\pm}$0.33	30.65$^{\pm}$0.03	36.10$^{\pm}$2.61	31.15$^{\pm}$0.91
4	28.50$^{\pm}$1.42	30.74$^{\pm}$0.28	36.03$^{\pm}$1.22	31.44$^{\pm}$1.53

F: significant (P=001) among treatments, within the same sampling.
not significant among sampling within the same treatment.

Newman and Keuls' Test (P=0.05)

Time 0	A	B	D	C
2 months	A	B	D	C
4 months	A	B	D	C

Table 5 reports data concerning the mineralisation of organic Nitrogen in field experiments.

Table 5. Mineralisation of organic Nitrogen in soils treated with sludge stabilized by different methods.
(Organic Nitrogen contents are reported as mg.Kg^{-1} of dry soil)

Treatments	Time 0		2 Months			4 Months		
	Total	Applicated	Total	% Mineralized Total	Applicated	Total	% Mineralized Total	Applicated
Control	3,000 $^{\pm}$3	–	2,997 $^{\pm}$132	0.12	–	2,888 $^{\pm}$161	3.99	–
Liquid sludge	3,146 $^{\pm}$129	76.49	3,140 $^{\pm}$27	0.18	7.33	2,991 $^{\pm}$157	4.92	202.13
Composted with wood chips	3,232 $^{\pm}$172	159.29	3,226 $^{\pm}$57	0.19	3.79	3,145 $^{\pm}$30	2.70	54.77
Stabilized with inerts	3,566 $^{\pm}$74	660.84	3,538 $^{\pm}$128	0.80	4.30	3,343 $^{\pm}$100	6.28	33.88

In this trial, the liquid digested sludge has shown to promote an intense mineralisation activity which, four months after the application, involves very large amounts of organic Nitrogen (about 200% of the applicated). The rather slow mineralisability of the organic Nitrogen of composted sewage sludges is quite confirmed. In all conditions the process appears faster in the last period of observation, probably as a consequence of higher environmental temperatures in addition to overcoming an eventual lag-phase. In table 6 the results of inorganic (ammoniacal and nitric) Nitrogen determina tions in field experiments are reported. They appear rather difficult to explain considering the influence of losses due to leaching, volatilization a.s.o.

Table 6. Mineral Nitrogen ($NH_4^+ + NO_3^-$, mg.Kg^{-1}of dry soil) in soils treated with sludges stabilized by different methods

Treatments		Time 0	2 Months	4 Months
Control	(A)	6.28±1.38	3.55±0.51	7.10±0.87
Liquid sludge	(B)	21.68±0.81	5.61±0.17	15.53±1.82
Sludge composted with wood chips	(C)	14.92±1.07	6.03±1.58	17.84±1.86
Sludge stabilized with inerts	(D)	107.81±11.0	28.40±0.97	14.60±1.16

F : significant at 1% level among treatments and sampling

Newman and Keuls' Test (P=0.05)

Time 0	A	C	B	D	
2 Months	A	B	C	D	
4 Months	A	D	B	C	

Variations of inorganic Nitrogen contents during incubation in laboratory experiments are resulted as in figures 2-4, where treatments are presented by the same symbols of figure 1. The largest amounts of inorganic Nitrogen have been detected in samples treated with liquid sludge, after two weeks of incubation and following a phase of organication. The treatments with composted sludges are responsible for not significantly different production of mineral Nitrogen, considering both application rate and type of sludge. All the differences resulted significant in comparison with the control. Initial and final data collected during the experiment as well as their statistical evaluation are reported in table 7. Figures 3 and 4 show that the production of nitrate must be considered responsible for the effect of the treatments, since unsignificant differences in ammonia production are resulted during the incubation.

Soil microbial biomass, quantified when soil respiration reached a linear trend, is reported in table 8 with the values of mineralisation coefficients (K_c) obtained according to Chaussod and Nicolardot (3) and utilized for biomass calculation. In the field, microbial biomass is just resulted greater in the soil treated with sludge composted in mixture with wood chips. On the other hand this soil sample showed a significantly lower K_c, as an expression of a different efficiency of organic carbon mineralisation, depending more on quality than on quantity of the organic matter applicated to soil. In laboratory experiments soil biomass appears significantly greater in the soil treated with liquid digested sludge. In this trials, lack of nutrients leaching has probably allowed a partial organication, resulting in the production of a really greater biomass. The mineralisation coefficient of control soil is not resulted much affected by incubation conditions. Its value, higher than other usually reported in the literature,

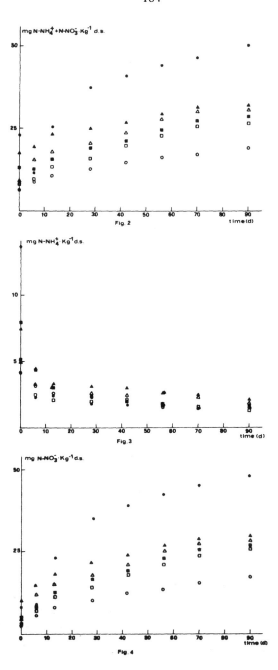

Fig. 2

Fig. 3

Fig. 4

has been confirmed in comparison to previous investigations carried on the same type of soil (5).

Table 7. Mineralisation of organic Nitrogen in soils treated with sludges stabilized by different methods. (Incubation trials) mg of Nitrogen . Kg^{-1} of dry soil are expressed

Treatments		Organic		Inorganic[*]	
		Applicated	Total after incubation	before incubation	after incubation
Control	(A)	–	3,161.02	6.28	18.80
Liquid sludge	(B)	76.49	3,237.16	22.83	50.25
Sludge composted with wood chips Rate I	(C)	39.82	3,200.84	7.96	24.60
Sludge composted with wood chips Rate II	(D)	159.29	3,320.31	12.99	28.47
Sludge stabilized with inerts Rate I	(E)	164.21	3,325.23	9.07	30.45
Sludge stabilized with inerts Rate II	(F)	660.84	3,821.86	17.44	31.80

[*]F: significant at 1% level

Newman and Keuls' Test (P=0.05)

A   C   D   E   F   B

FDAasic activity has been measured in the field 45 and 110 days after the treatments; in laboratory experiments, after 15, 60 and 98 days of incubation. In field, the highest value coincided with the greatest biomass at the end of the experiment. It was in fact detected in the soil treated with sludge composted in mixture with wood chips. In vitro, no significant difference resulted after 15 days of incubation; but after 60 days the activity of soils treated with lower doses of compost or with liquid sludge was indeed higher. The same difference resulted after 98 days too. All these data are reported in table 9. They allow to conclude that sludge applications stimulate soil hydrolytic activities; but, at the end of the study, these activities are not proportional to the application rate, because the most favourable conditions have been pointed out for samples treated with the lowest amounts of organic material. Sludge quality doesn't seem to exert significant influence.

Microbiological analyses of soil samples, when mineralisation process reached a linear trend, have provided the results of table 10. Means comparison, carried by Student's test after logarithmic data normalisation, has put into evidence that treatments have not disturbed soil microbial populations. Ammonifier microorganisms are significantly and weakly depressed in

Table 8. Microbial biomass (mg of $C.Kg^{-1}$ of dry soil) in soils treated with sludges stabilized by different methods

Treatments		Incubation experiments		field experiments	
		Biomass	Kc	Biomass	Kc
Control	(A)	231.58±49.9	0.63±0.06	331.56±11.97	0.61±0.01
Liquid sludge	(B)	418.58±36.40	0.53±0.06	348.95±29.38	0.63±0.04
Sludge composted with wood chips Rate I	(C)	267.80±36.09	0.62±0.04		
Sludge composted with wood chips Rate II	(D)	313.97±18.94	0.63±0.02	733.50±28.83	0.55±0.03
Sludge stabilized with inerts Rate I	(E)	301.44±29.96	0.56±0.03		
Sludge stabilized with inerts Rate II	(F)	353.56±23.60	0.55±0.03	372.98±34.08	0.62±0.04

F: significant at 1% level

Newman and Keuls' Test (P=0.05)

Biomass	A	C	E	D	F	B		A	B	F	D

Kc	B	F	E	C	D=A		D	A	F	B

soil treated with liquid sludge only. NH_4^+-oxidizers don't seem significantly affected. NO_2^--oxidizers exhibit higher counts in soils treated with liquid sludge or with sludge stabilized in mixture with inert bulking agents.

4. CONCLUSIONS

Technical troubles of a study on mineralisation of an organic material after its application to soil are well known. It is also known that attempt to obtain a correct balanceof the various products arising from the transformation of organic matter within a complex natural ecosystem like the soil inevitably gives rise to experimental circumstances which are often very different from natural conditions. At a same extent, it is finally known that results from every experimental effort, in this type of study, must be fundamentally considered valid for the conditions utilized only.

The main aim of the research reported in this paper was to point out, through field and laboratory investigations, possible differences among the mineralisation rates of organic matter contained in a sewage sludge stabilized by different methods, and to evaluate effects on some microbiological soil characteristics.

As far as organic Carbon mineralisation in the field, results of analyses have only reflected the amounts applicated. In vitro, the process

is resulted faster for liquid digested sludge. The application of this type
of material to a soil with a poor organic matter content, has shown a
positive priming effect, causing the mineralisation of an organic Carbon
amount greater than the applicated one. Different mineralisation kinetics
have been defined for the various sludges assayed. Moreover it has been
pointed out that the mineralisation of organic Carbon of sludges stabilized
in the solid phase is strongly affected by the application rates, as fas as
kinetics are concerned. The sludge composted with wood chips is significant
ly resulted able to increase the organic Carbon soil content.

Table 9. FDA hydrolytic activity (O.D.$_{490}$) in soils treated with sludges
stabilized by different methods

Treatments		Time (days)				
		15°	60$^{\#}$	98$^{\#}$	45$^{\#\#}$	110$^{\#\#}$
		Incubation experiments			Field experiments	
Control	(A)	0.18$\pm$0.01	0.43$\pm$0.07	0.24$\pm$0.02	0.52$\pm$0.06	0.42$\pm$0.08
Liquid sludge	(B)	0.17$\pm$0.03	0.84$\pm$0.06	0.86$\pm$0.06	0.77$\pm$0.11	0.40$\pm$0.05
Sludge composted with wood chips Rate I	(C)	0.15$\pm$0.03	0.82$\pm$0.07	0.94$\pm$0.11		
Sludge composted with wood chips Rate II	(D)	0.25$\pm$0.11	0.57$\pm$0.07	0.29$\pm$0.01	1.14$\pm$0.33	0.94$\pm$0.06
Sludge stabilized with inerts Rate I	(E)	0.19$\pm$0.04	0.84$\pm$0.10	0.93$\pm$0.05		
Sludge stabilized with inerts Rate II	(F)	0.17$\pm$0.02	0.41$\pm$0.06	0.26$\pm$0.07	0.87$\pm$0.31	0.77$\pm$0.06

F: o =not significant; $\#$ =significant at 5% level; $\#\#$ =at 1% level

Newman and Keuls' Test (P=0.05)

| 60 days | F | A | D | C | B=E |
| 98 days | A | F | D | B | E | C |

| 45 days | A | B | F | D |
| 110 days | B | A | F | D |

In our experimental conditions the application of liquid sludge has
also promoted an intense organic Nitrogen mineralisation, especially with
respect to application rate. This type of sludge, after a phase including
organication phenomena, makes available into the soil the largest amounts
of inorganic Nitrogen. The slower organic Nitrogen mineralisability of
composted sewage sludges has been confirmed, but the behaviour of the sludge
stabilized in mixture with inert bulking agents seems to be particularly
interesting. Indeed this material shows to be able to release some of its

Nitrogen within a short, then mineralizing more slowly.

Among treatments, the application of sludge composted in mixture with wood chips has caused in field significant increases of soil microbial bio mass and hydrolytic activities. In the incubation studies the highest values of biomass have been obtained in samples treated with liquid sludge.

The various treatments, proportioned in this study to organic matter content of materials, have not produced considerable effects on counts of ammonifier microorganisms, NH_4^+-oxidizers, and NO_2^--oxidizers, quantified when the mineralisation processes reached a linear trend. Effect of repeated applications must be nevertheless verified.

Table 10. Microbial counts (MPN of viable cells.g^{-1} of dry soil) in soils treated with sludges stabilized by different methods

Treatments		Ammonifiers $x10^6$	NH_4^+-oxidizers $x10^3$	NO_2^--oxidizers $x10^3$
Control	(A)	15.94	23.69	0.85
Liquid sludge	(B)	3.73	7.38	27.81
Sludge composted with wood chips Rate I	(C)	18.30	22.91	1.53
Sludge composted with wood chips Rate II	(D)	5.11	8.12	1.59
Sludge stabilized with inerts Rate I	(E)	18.20	9.23	2.64
Sludge stabilized with inerts Rate II	(F)	17.89	14.48	20.00

Means comparison (t-Student) after logarithmic normalisation (P=0.01)

Ammonifiers	B	D	F	A	E	C
NH_4^+-oxidizers	B	D	E	F	C	A
NO_2^--oxidizers	A	C	D	E	F	B

The three different sludges assayed in this research have altogether shown behaviours which cannot be neglected within a rational practice of agricultural utilization of sewage sludge.

AKNOWLEDGMENTS

This work has been supported by grants of Ministero per la Pubblica Istruzione and of Consiglio Nazionale delle Ricerche, Roma. Authors are also indebted with Mrs. Rosa Maria Andolfi for her technical co-operation.

REFERENCES

1) ALEXANDER, M. and F.E.CLARK (1965) - Nitrifying bacteria.pp 1477-1486. in C.A.BLACK et al. (Eds). Methods in soil analysis.Part II.American Soc.of Agronomy. Madison, Wisc.U.S.A.

2) COMMISSION OF THE EUROPEEN COMMUNITIES (1979) - Workshop on standardi zation of analytical methods for manure, soils, plant and water. Gent (Belgium), 6-9 June.

3) CHAUSSOD, R. and B.NICOLARDOT (1982) - Mesure de la biomasse microbien ne dans les sols cultivés.I.Approche cinétique et estimation simpli fiée du carbone facilement minéralisable. Rev.Ecol.Biol.Sol, 19, 501-512.

4) COPPOLA, S., DUMONTET, S. and P.MARINO (1983) - Composting raw sewage sludge in mixture with organic or inert bulking agents. International Conference on Composting of Solid Wastes and Slurries. University of Leeds (UK), 28-30 September.

5) DUMONTET, S. and E. PARENTE (1984) - Effets d'une boue de station d'épuration sur la microflore tellurique et sur ses activités. Ann.Fac.Sc.Agr.Univ.Napoli, IV, 18, 9-36.

6) POCHON, J. and P.TARDIEUX (1962) - Techniques d'analyse en microbiolo gie du sol. Ed.de la Tourelle, St.Mandé (France).

7) SCHMÜRER, J. and T.ROSSWALL (1982) - Fluorescein diacetate hydrolysis as a measure of total microbial activity in soil and litter. Appl. Environ.Microbiol., 43, 1256-1261.

8) SNELL, F.D. and C.T.SNELL (1961) - Colorimetric methods of analysis. Vol. II, Van Norstrand Co., Princeton.

9) SORIANO, S. and M.WALKER (1968) - Isolation of ammonia-oxidizing autotrophic bacteria. J.Appl.Bacteriol., 31, 493-497.

10) STRICKLAND, J.D.H. and T.R.PARSON (1968) - A practical handbook for sea-water analysis. Fisheries Research Board of Canada.Bull.n.167.

SLUDGE ORIGINS AND NITROGEN EFFICIENCY

L. BARIDEAU[x] and R. IMPENS[xx]

(x)   Groupe Valorisation des boues, Faculté des Sciences agrono-
      miques, 5800 Gembloux.
(xx) Professor, Chaire de Biologie végétale, Faculté des Sciences
      agronomiques de l'Etat, 5800 Gembloux.

Summary

The effect of sludges produced in three very different sewage plants
on the growth of ray-grass (Lolium multiflorum L.) at two temperatures and four sludge application rates was tested. The existence of
toxic products in the sludges was tested before the trial by a germination test with cress seeds (Lepidium sativum L.). One of the
sludges inhibited strongly the germination of cress seeds, but this
effect didn't affect the growth of the ray-grass during the trial.
Dry matter production and nitrogen exportation were measured. Dry
matter production was multiplied by a factor ranging between 1,05 and
2,50 and nitrogen exportation by a factor ranging between 1,1 and
2,70 as compared to the control. Temperature increased the production
by 3 to 20 % and nitrogen exportation by 9 to 33 %.
The nitrogen efficiency was computed and found to be independant of
temperature but largely dependant of the sludge origin.

1. INTRODUCTION

    The agricultural value of the sludges is mainly linked to their ni-
trogen content and availability. Fürrer and Bolliger and others have
shown that nitrogen efficiency was related to the forms of nitrogen in the
sludge, the ammonium form being almost entirely available and the organic
form being less available. However, the importance of climatic conditions
on the availability of nitrogen can't be underestimated : temperature and
moisture are important factors in that respect. The origin of the sewage
and the methods of treatment and sludge stabilization are certainly another important factor of nitrogen availability. In this trial we have compared the effect of these two factors, sludge origins and climatic conditions, on nitrogen availability.

2. Methods

2.1. Origin and composition of the sludges

    Three sludges have been selected, coming from plants with very diffe-
rent characteristics : Rhisnes, Wasmuel and Hannut.
    The sludges of Rhisnes and Hannut are commonly used by farmers.
Wasmuel produced toxic sludges some years ago and since then the sludges
have never been used in agriculture. The three plants possess the usual
features of primary and secondary settling tanks. Their other characteristics can be described has follows :

- <u>Rhisnes</u> : the waters treated by this plant come from a cheese factory. The sewage flows through a bacterial bed and the sludges are aerobically stabilized. The plant has a capacity of 11500 EH.

- <u>Wasmuel</u> : this plant receives the sewage of the city of Mons and from some neighbouring industries. The methods of treatment here are aerated sludge and anaerobical digestion. The capacity of Wasmuel is 225.000 EH.

- <u>Hannut</u> : Hannut is a typical small rural town. Its sewage plant is fitted up with a bacterial bed and a cold anaerobical digestor. The capacity of this plant is 5300 EH.

The liquid sludges were sampled in november 1983. Their dry matter and total nitrogen content appear in table 1.

Origin	D M (%)	total N (% D M)
Rhisnes	6,69	3,54
Wasmuel	5,59	2,93
Hannut	7,39	2,59

Table 1 : Dry matter and total nitrogen content of the three sludges.

2.2. Growth conditions

Italian ray-grass (Lolium multiflorum L.) was selected for this trial. It was grown in two different climatic conditions summarized in table 2. The high level of temperature was obtained in a conditioned room, with artificial lightning ; the low level in a warmed greenhouse with lightning in the morning and the evening to increase the photo period to 12 hours. In both cases Radium HRI-T 400 W/DH lamps were used.

The pots used were black plastic containers of 13 x 13 x 13 cm filled with 1,5 kg of soil. The soil is the top layer of a typical agricultural soil from the Gembloux region (sandy loam).

Ray-grass has been sown some days after the mixing of soil and sludges at the rate of 1,2 gr of seeds per pot.

Level	Temperature average		Lightning	Photoperiod	Relative humidity
	Day	Night			
High	21	18	artificial	12 + 12	40 - 60
Low	13	13	natural + artificial	12 + 12	60 - 70

Table 2 : Growth conditions for the two selected growth conditions : temperatures in °C and photo periods in hours, relative humidity in %.

2.3. Sludge rates applied

The sludges are compared on the basis of the amount of dry matter applied to the soil and not on the volume of sludge applied. We feel that this gives a better comparison between sludges at least when the dry matter contents are not too different. The sludges are applied at four rates : 0; 2,5; 5 ; 10 gr DM per pot, this corresponds roughly to 30, 60 and 100 m^3/Ha as far as the extrapolation is possible. Table 3 gives the quantities of sludge and the corresponding amounts of nitrogen brought with them.

We used a split-plot design in both locations with the origins as great plots and the sludge rates as small plots, with four replications.

Origin	Sludge quantities			total Nitrogen		
Rhisnes	37	75	149	89	177	354
Wasmuel	45	89	179	73	147	293
Hannut	34	68	135	65	130	259

Table 3 : Sludge (ml) and corresponding total nitrogen
(mg) applied per pot.

3. Preliminary trial

The Wasmuel sludges being reputed toxic to the plants we compared the three sludges in a germination test before starting the pot experiment. Although this test falls somewhat out of the scope of the purpose of the trial, we think it is interesting to give its results here and to compare them with those obtained subsequently in the pot trial.

3.1. Germination test conditions

Sludges were mixed to fine sand in Petri dishes. The proportions used were pure sand ; 20 % sludge – 80 % sand ; 50 % sludge – 50 % sand ; sludge only. The mixture was covered with filter paper (Whatman n° 1), and 100 cress seeds (Lepidium sativum L.) spread on the paper. Water was added to maintain a suitable humidity when necessary. Each treatment was repli-cated five times. The dishes were maintained at 25 ° C and the number of germinated seeds were counted after 36 hours. The results expressed as germination inhibition percentage are given in table 4. A more synthetic expression of these results is given as ID 50 (dose that inhibits the germination of 50 % of the seeds). This result was obtained graphically by plotting the results on Probit scale.

Sludge rates	Origins		
	Rhisnes	Wasmuel	Hannut
0	12,2	12,8	10,6
20	9,6	83,2	23,2
50	68,0	98,2	83,0
100	98,4	99,0	89,4
ID 50	40	10	56

Table 4 : Germination inhibition per rates
for the three sludges (%) and
ID 50 per origin (%)

3.2. Conclusion of the germination test

The test seemed to confirm the existence of some toxic properties of the Wasmuel sludge and their absence in the Rhisnes and Hannut sludges. Nevertheless we decided to go on with the trial as scheduled. We will see hereafter how different the results of the pot trial are from these of this preliminary trial.

4. Results of the pot trials

4.1. Dry matter production

The ray-grass has been harvested after 34, 70, 120 and 175 days after sowing. Growth has ceased in the low temperature condition, but is still going on in the high temperature condition : a fifth cut has been realized and should accordingly modify the results obtained as far. We present in table 5 the total amount of dry matter obtained for the four cuts in the two different temperature conditions.

Origin	Sludge	Temperature				
		High		Low		
	Dry matter (g)	Dry matter (g)	%	Dry matter (g)	%	
Rhisnes	2,5	3,514	137,1	3,231	149,9	
	5	4,763	185,8	4,338	201,2	
	10	4,569	178,2	4,878	226,3	
Wasmuel	2,5	3,827	149,3	3,151	146,2	
	5	5,138	200,4	4,304	199,6	
	10	5,424	211,6	5,423	251,5	
Hannut	2,5	2,702	105,4	2,659	123,3	
	5	3,145	122,7	3,085	143,1	
	10	3,832	149,5	2,611	121,1	
Control	0	2,564	100,0	2,156	100,0	

Table 5 : Total and relative production per pot
in the two growth conditions.

A global statistical analysis of the yiel data shows that the differences between growth conditions, sludge origins and sludge rates are
highly or very highly significant, that very highly significant interaction between sludge origins and rates and a highly significant interaction between the three main factors exist. However interactions between
sludge origins and temperature or sludge rates and growth condition are
not significant. In both growth conditions, the Hannut sludge gave significantly lower yields than the two other sludges.

The increase in production obtained by the so-called toxic Wasmuel
sludge is quite astonishing and far from what could be expected by the
germination trials. The disappearance of the Wasmuel sludge toxicity can
be explained by the dilution of the sludge into the soil and by the ad-
sorption of the germination inhibition factor on the soils colloids and
organic matter. Whatever this inhibition factor was, it didn't affect the
growth of the ray-grass anymore. The Wasmuel sludge increases the yield
by amounts ranging between 50 and 150 % of the controls in both environ-
ments. The Rhisnes sludge increases the yields by slighly smaller amounts
than the previous sludge.

The influence of temperature on the dry matter production can be
emphasized by expressing the mean yield per origin at high temperature
level as a percentage of the mean yield at low temperature (table 6). It
appears thus that temperature had an important effect on the control and
on the sludge of Hannut, a slighly smaller effect on the yield of that of
Wasmuel and almost no effect on the yield of the Rhisnes sludge.

Origin	Temperature	
	High	Low
Rhisnes	103,2	100,0
Wasmuel	111,7	100,0
Hannut	115,8	100,0
Control	118,9	100,0

Table 6 : Mean yields per origin as a percentage of low temperature yields.

4.2. Nitrogen exportation

The nitrogen exportation was obtained by multiplying the dry matter production per pot by its nitrogen content. The results are given in table 7. The differences between the different factors are obvious, and the statistical analysis confirms this : differences between growth conditions, sludges origins, sludge rate, and the origins x rates interaction are very highly significant ; the interaction of the three main factors is significant.

Sludge origin	Sludge rate (g)	Temperature			
		High		Low	
		exported N	%	exported N	%
Rhisnes	2,5	95,5	136,2	75,5	140,3
	5	120,3	171,6	108,8	202,2
	10	120,0	171,2	122,7	228,1
Wasmuel	2,5	103,0	146,9	83,5	155,2
	5	145,7	207,8	127,2	236,4
	10	179,0	255,3	146,0	271,4
Hannut	2,5	77,5	110,6	67,5	125,5
	5	91,2	130,1	81,7	151,9
	10	124,0	176,9	70,5	131,0
Control	0	70,1	100,0	53,8	100,0

Table 7 : Exported nitrogen (mg) per pot in the two growth conditions

At low temperature, the three origins are significantly different
from each other, but at high temperature, Rhisnes and Hannut sludges do
not differ from each other and are significantly lower than the Wasmuel
sludge.

Just as for the yield results, we can express the mean nitrogen
exportation per origin at high temperature as a percentage of the mean
exportation at low temperature (table 8). It will be noted that these results are quite well correlated with the yields results, but that the temperature has a higher influence on nitrogen exportation than on yield.

Origin	Temperature	
	High	Low
Rhisnes	109,4	100,0
Wasmuel	119,9	100,0
Hannut	133,2	100,0
Control	130,2	100,0

Table 8 : Mean nitrogen exportation
per origin as a percentage
of low temperature exportation

5. Nitrogen efficiency

The nitrogen exportation data allows us to compute the efficiency of
the sludge nitrogen : if the high temperature conditions increased the
organic matter mineralization and release of nitrogen, the efficiency of
the sludge nitrogen should be higher in the high temperature conditions.
If we assume that the sludges had no influence on the soil organic matter
mineralization and on soil nitrogen dynamics, or in other words if the
sludge effects are additive, then the difference between the nitrogen exported by the yields of the sludge treated pots and that exported by the
yields of the control plots represents the nitrogen supplied by the
sludges.

The nitrogen efficiency can then be defined as :

$$E = ((N_t - N_c)/N_s) \times 100$$

with N_t : nitrogen exported by the yield of a sludge treated pot
N_c : nitrogen exported by the yield of the control
N_s : nitrogen imported by the applied sludge

The results of these computations are given in tables 9 and 10.

Sludge origin	Sludge dry matter	High T°	Low T°
Rhisnes	2,5	28,6	24,3
	5	28,3	31,0
	10	14,1	19,5
Wasmuel	2,5	45,1	40,6
	5	51,5	50,3
	10	37,2	31,5
Hannut	2,5	11,5	21,0
	5	16,0	21,5
	10	2U,8	6,4

Table 9 : Nitrogen efficiency (%) for the
different treatments

5.1. Comments on the nitrogen efficiency results

The temperature did not affect the average nitrogen efficiency but
the reactions of the different sludges to the increase of temperature are
somewhat different : the nitrogen efficiency decreased slightly for Rhisnes
and increased for Wasmuel and Hannut.

Differences between sludge origins are more important as can be seen
in table 10 : the nitrogen supplied by the Wasmuel sludge is at least
twice more efficient than the nitrogen from the two other plants. If this
is the effect of a general law on sludge nitrogen availability, one should
be allowed to concluded that aerated sludge and anaerobical digestion provides the best sludge when nitrogen availability is considered.

Sludge origins	High T°	Low T°	Average
Rhisnes	20,2	23,5	21,9
Wasmuel	42,4	38,2	40,3
Hannut	18,2	12,9	15,5
Average	26,8	25,2	26,0

Table 10 : Nitrogen efficiency per origin
and for the two growth condi-
tions (%)

The sludges treated by other methods, in this case bacterial beds and aerated stabilization or cold anaerobical digestion, seems to contain less efficient nitrogen, even if their nitrogen content is higher than that of the aerated-anaerobicaly digested sludge (table 3).

Another explanation should be that the stabilization processes of Rhisnes and Hannut do not work properly, and that this has an important effect on sludge nitrogen availability. If the hypothesis can be accepted in the case of Hannut, it is difficult to accept it for Rhisnes, where many controls are effected by the people in charge of the plant. The initial hypothesis should then have some consistency.

6. Conclusion

As expected, temperature had a favourable influence on the dry matter production of the ray-grass and on its nitrogen content : this can be explained by an increase of organic matter mineralization due to the increase of temperature. Temperature did not influence the nitrogen availability, as could be expected : the proportion of sludge nitrogen exported by the ray-grass remains practically the same at high or low temperature. Nitrogen availability seems to be strongly dependant of the sludge origin, and is probably correlated with the nature of sewage treatment.

7. Acknowledgments

This work was part of a project on information of the farmers on the use of sewage sludge in agriculture and was granted by the minister of walloonian region for water, environment and rural life. The authors are indebted to this minister for the aid provided. They want to thank Mrs Cambier and MMrs Dekeirsschieter, Doucet and Gigot for their technical assistance.

8. Bibliography

Barideau L. et Falisse A. La disponibilité de l'azote dans les boues résiduaires. Bull. Rech. Agron. Gembloux (1982), 17(3), 227-236.

Fürrer O.J. und Bolliger P. (1978) Die Wirksamkeit der Stickstoffes in Klärschlam Schweiz. Landwirtsch. Forsch. 17 (3-4), 137-147.

Hanotiaux G., Heck J.P., Rocher M., Barideau L. et Marlier-Geets O. (1980). Evolution of phosphorus in sewage sludge after its application to the soil. In : Phosphorus in sewage sludge and animal waste slurries. Proceedings of an EEC Seminar (Groningen). Dordrecht, D. Reidel Publishing Company, 399-410.

Heck J.P., Barthélémy J.P. et Rocher M. (1978). Caractérisation de certains éléments biogènes majeurs et métaux lourds dans les boues de stations d'épuration d'eaux usées. Ann. Gembloux 84, 77-92.

Heck J.P., Louppe L., Marlier-Geets O., Rocher M. et Barideau L. (1980). Evolution dans le sol des différentes formes d'azote présentes dans les boues. In : Characterization, treatment and use of sewage sludge. Proceedings of the second European Symposium (Vienna). Dordrecht, D. Reidel Publishing Company, 466-474.

Xanthoulis D. (1980). Valorisation agricole des boues d'épuration en Wallonie. Ann. Gembloux 86, 61-77.

Xanthoulis D. et Falisse A. (1978). Utilisation des boues résiduaires en grandes cultures. Ann. Gembloux 84, 101-109.

DISCUSSION

Chairman: Dr O J Furrer, Switzerland

J VOORBURG to E VIGERUST

In your last slide, you showed the effect of composting sewage sludge.
Are the differences in crop growth explained by differences in available N
or are there other effects?

Answer: The differences in growth can mainly be explained by the lower
nitrogen effect as the sludge is more decomposed. From raw uncomposted
sludge phytotoxic effects are common.

S DE HAAN to E VIGERUST

Do I understand that in Norway you are advocating storage of the sludge
if it is to be used as a soil conditioner? It is better to use fresh
sludge if it is required for its nitrogen value.

Answer: In Norway, we favour composting as the process reduces toxicity
effects of the sludge. If it is composted for too long a period of time
it is not such a good soil conditioner.

S DE HAAN to E VIGERUST

What steps are taken to prevent leaching during storage for composting?

Answer: Sites for sludge storage have to be carefully considered with
regards to pollution. In summer the leachates will infiltrate in the
soil. In winter the sludge surface will be frozen and there will be very
little or no biological activity and no leaching.

F STADELMANN to S COPPOLA

How do you explain the high biomass in the field experiments in the case of
sludge composted with wood chips compared with the biomass of the same
sludge in the incubation experiments? The biomass results of incubation
and field experiments in case of liquid sludge and sludge composted with
wood chips do not agree.

Answer: Values in the two trials cannot be compared because of the
differences between the amounts of material applied and between the
experimental conditions in each case. In the case of sludge composted with
wood chips, in the incubation experiments we have an inferior biomass with
a higher turnover; it is the contrary in field experiments.

O FURRER to S COPPOLA

What was the ratio of dry matter from wood chips and from sludge?

Answer: It was 4 parts sludge (14% dry matter) to 1 part of wood chips by weight.

J WILLIAMS to S COPPOLA

Mineralisation appears to be strongly affected by application rate, is this likely to be due to the heavy metal concentrations achieved in soil, eg Hg, which could have a fungitoxic effect?

Answer: This aspect has been looked at in previous work and the more likely explanation was the presence of some phytotoxin developed in the compost and not the heavy metal effect.

S de HAAN to S COPPOLA

I have two questions concerning table 1 in your paper:-

1. The C content of the liquid digested sludge can be calculated as 64% and as 21% for the sludge composted in mixtures with wood chips or inert agents. As far as I know the C content of organic soil amendments is never less than 40% of OM (sugar) and never higher than 60% (humus 50%). Can you explain the high value for your liquid digested sludge and the low value for the composted sludges?

2. The C/N ratio of your liquid digested sludge is about 16, which can be regarded as quite normal. Following mixing the sludge with wood chips (with normally a very large C/N ratio) one would expect an increase in the C/N ratio but in your sludge composted with wood chips, the contrary is the case. Can you explain this, and especially can you explain the very low C/N ratio (about 3) of your sludge composted in mixture with inert agents? What was the type of the inert agent? Did it contain nitrogen?

Answer: I do not have the facts to reply to the first question. On the second question, I must say that C/N ratios are not achieved through the composting processes as performed in our laboratory. Composting in mixture with wood chips, raw sewage sludges, dewatered up to about 14-15 per cent of dry matter (4 parts by weight) are mixed with wood chips (1 part). The product is treated twice in the same manner and finally screened to eliminate the residual largest parts of wood chips, so that, at the end, the wood contribution to the compost composition is the smallest. Moreover, organic carbon mineralisation during the composting process must be taken into account.

Composting raw sludge in mixture with inert bulking agents only involves the sludge material, as the bulking agents, chemically constituted by polyolephinic polymer, do not participate in the reaction. The negligible losses of nitrogen during this type of sludge stabilization, especially in comparison with other composting methods, have already been put into evidence.

B POMMEL to L BARIDEAU

You observed differences in nitrogen efficiency depending on rate of sludge
application, can you offer any explanation for this difference?

Answer: I think that the quantity of water brought with the higher rate
of sludge played a role in this phenomenon: the substrate was compacted
and this certainly caused a problem for root growth. But this is only the
physical effect and does not explain everything. Some effect linked with
nitrogen metabolism in the soil is most probably involved like leaching or
denitrification, but I didn't investigate this.

J HALL to L BARIDEAU

In the UK we have identified a sludge which failed anaerobic digestion and
was therefore considered unsuitable for agricultural use yet in a
comparable test to yours, it produced by far the highest yields of grass.
Have you identified the toxin responsible for inhibiting germination and
secondly what were the ammonia contents of the sludges?

Answer: The germination inhibition factor was not identified in this case,
but it appeared some years ago that quaternary ammonium compounds
intoxicated the biomass of the aerobic treatment process of the sewage
plant and caused toxicity phenomena in the sewage sludge. It is thus
possible that these compounds were still present in the waste waters
treated and that they are responsible for the germination inhibition.
Unfortunately, we didn't measure the ammonium nitrogen content of the
sludges. This result may have provided an answer to the quaternary
ammonium hypothesis.

C. METHODS FOR EVALUATING THE COMPOSITION OF

SLURRIES AND MANURES

Evaluation of type and contents of humic substances in sludges and composts

Evaluation of urban and animal wastes as sources
of phosphorus

Slurry-meter for estimating dry matter and nutrient content of slurry

Discussion

EVALUATION OF TYPE AND CONTENTS OF HUMIC SUBSTANCES
IN SLUDGES AND COMPOSTS

M. DE NOBILI, G. CERCIGNANI, L. LEITA

Istituto di Produzione Vegetale dell'Università di Udine

Summary

Criteria for evaluating the humification degree of organic amendments are pratically lacking.

Humic substances content and quality have been investigated in samples from sewage sludge and composting plants. By determining total organic carbon in $Na_4P_2O_7$ extracts and in fractions not adsorbed on Polyclar AT, the ratio of non humic to humic substances was calculated. This ratio may be considered an index of the stabilization and possible low toxicity of the material.

The quality of humic substances from a number of different organic wastes has been examined by isoelectric focusing (IEF) and compared to that of stabilized soil organic matter. Humic acids from fresh and mature farmyard manure and from worm compost, show great similarity with soil humic acids, while samples of anaerobically digested sewage sludge or immature compost were seen to have completely different patterns. Maturation of compost or permanence in thickening beds lessens the differences in IEF patterns relative to stabilized materials.

1. INTRODUCTION

Recycling of urban and industrial wastes for agricultural land fertilization needs a suitable control of these materials before field application. Recently, a draft proposal for compost specifications has been presented to the Commission of European Communities by Zucconi and De Bertoldi (1); according to these Authors, compost is defined as "a self stabilized and sanitized product of controlled bioxidative degradation of heterogeneous organic matter in a solid state (composting)." Compost should be as "a phytoergonic product which undergoes a slow humification process (stabilization), having past the fast metabolic stage of initial decomposition ".

Central to this definition is the concept of "stabilization" which is to be reached before use, in order to avoid phytotoxic effects; this equally applies to products deriving from maturation of sewage sludge and farm manure. To date, however, a clear cut working definition of "stabilization" for composting materials and other organic wastes is still lacking. Among the chemical parameters required for the characterization of the stabilized product of organic wastes, is the amount and humification degree of organic matter (1). The classic analysis of humified organic matter (HOM), accomplished by determining humic (HA) acid and fulvic (FA) acid content (2) is not likely to be satisfactory in this context. Many different kinds of substances can be solubilized and co-extracted, together with humic compounds, and determined as HA or FA, simply on the basis of their solubility in water at pH=2.

We therefore sought a suitable procedure to detect any variation in humic substance content and quality in sewage sludges manure and composts. Soil FA are phenolic-type compounds, they are known to be completely adsorbed on Polyclar AT at pH less than two: while coloured materials of soil FA are quantitatively adsorbed on Polyclar at pH=2, the same is not true for the "FA" fraction prepared from different materials. Organic substances not adsorbed on Polyclar can be considered non-stabilized, and a possible source of toxicity after application; their quantification is therefore of importance from this point of view. On the other hand, more information is needed about the kind of humic substances present in organic wastes. Consequently, we have also used IEF as a useful mean of direct insight and qualitative comparison among humic substances obtained from commonly used organic wastess, soils, composts and sewage sludge.

1. <u>MATERIALS AND METHODS</u>

Sewage sludge samples were collected from the municipal sewage depuration plant in Udine. They were of three types: sludge from the anaerobic digestor (US1), freshly laid-down sludge (US2), and four-months-old sludge (US3) from thickening beds. Compost samples from Udine municipal composting plant were collected at three different stages: at the outlet of the plant (UC1), during maturation (UC2), and mature compost as delivered to farmers (UC3). Two more compost samples (finished products) were obtained from the municipal composting plants in Lignano (Udine) (L) and Pistoia (P). Other organic waste samples for analytical IEF were collected at different locations in Friuli (Italy), while earthworm compost samples (finished product) were obtained by local producers.

Soils, manures and composts were oven-dried at 105°C. After extraction with 0.1 M $Na_4P_2O_7$ pH=7.1 for 24 hours at 37°C, samples were centrifuged at 4000 rpm for 20 minutes and the supernatants filtered through a 0.2 microns membrane. HA was precipitated by addition of conc. H_2SO_4 and separated by centrifugation; the supernatant was passed twice through a column (1x4 cm) filled with polyvinylpyrrolidone (Poly-clar AT, SERVA). The non-retained fraction was collected and analyzed for total organic carbon. FA fraction was eluted from the column with 0.5 M NaOH and added to redissolved HA before treatment with IR 122 (SERVA, acid form) and neutralization (HA+FA fraction).

IEF runs were performed on HA+FA fractions as described by Ceccanti et al. (3); carrier ampholytes (range: 3.5 to 10) were purchased from LKB (Sweden). Organic carbon determinations were carried out on 1-g (solid) or 5-ml aqueous samples) aliquots by a chromic acid titration method (4), using a Mettler Memo Titrator DL 40 RC apparatus.

3. RESULTS

Humified organic matter. Table 1 reports data on the organic car-bon content of fractions from sewage sludges and composts. As can be seen, not only does the amount of organic matter vary among different samples, but the fraction not adsorbed on Polyclar after acidification is also varying. Humified organic matter, which is adsorbed on Poly-clar, shows an absolute increase during maturation of sewage sludge, but this seems not to be true in the case of compost.

Analytical IEF. As can be seen from Fig. 1, the IEF patterns of humic substances extracted from soil, fresh and mature farmyard manure, poultry manure and worm compost are strikingly similar to each other and are all characterized by a great heterogeneity of bands in the pH gradient region from 6.5 to 4.5 (5). Compost samples collected from Udine composting plant, on the other hand, show a simpler pattern, expecially in samples taken from the outlet of the processing plant or from freshly laid heaps, whose coloured bands of humic substances focalize only at pH values lower than 5.5. The older sample (UC3), how-ever, displays bands with less acidic isoelectric points, and a tenden-cy to develop an IEF pattern more similar to that of soil organic mat-ter and other well humified materials. According to qualitative evalua-tion by IEF, compost samples from Pistoia and Lignano composting plants are more efficiently humified in this respect. Udine sewage sludge sam-ples are characterized by a more complex composition than Udine com-post, also in the sample from the anaerobic digestor. Permanence in the thickening beds, and the consequent going on of humification process causes a progressive strengthening of bands in the pH region between 6.5 and 5.5.   (Fig.2 and 3).

4. DISCUSSION

A definition of the stabilization and humification degree for an organic amendment can be reached only by taking into account several parameters. In the present work, we are dealing with some qualitative and quantitative specifications of organic wastes. Our results show that both parameters are relevant to the evaluation of a maturation process. As quantitative analysis is concerned, we propose a tentative index of the humification degree, based on the ratio between non-humified organic matter (not adsorbed on Polyclar) and humified organic matter. Of course, lower values are expected for matured materials; actually, very low values are found for soil organic matter. From the data in Table I, the following values for this ratio are obtained:

US1	0.76
US2	1.20
US3	0.70
UC1	1.69
UC3	0.55
L	0.13
P	1.67

Analytical IEF provides a qualitative approach to the description of the maturation process: we suggest that stabilized organic wastes should have an IEF pattern similar, as much as possible, to that of organic matter of soil or materials of natural origin, such as farmyard and poultry manure, worm compost and so on. Sewage sludges show a lowering of the above proposed ratio during maturation (sample US2 and US3); the US1 sample may be considered anomalous, since it was collected during the anaerobic digestion. The IEF patterns agree with these results. Among the various compost samples, the Pistoia sample displays a good similarity to soil organic matter in IEF, whilst the proposed ratio is not low; this could be due to the high carbon content of the P compost. The reported findings concerning the remaining compost samples are in agreement with the proposed criteria.

ACKNOWLEDGEMENT

Research carried out with a grant from the Italian Ministry of Public Education.

REFERENCES

1. F. Zucconi, M. De Bertoldi. Draft proposal for Compost Specifications R&D Programme Recycling of Urban and Industrial Waste. Commission of European Communities (1984).

2. M. Schnitzer, L.E. Lowe, J.F. Dormaar, Y. Martel. A procedure for the characterization of soil organic matter. Can. J. Soil. Sci. 61, 517-519 (1981).

3. B. Ceccanti, P. Nannipieri, M. T. Bertolucci. Characterization of soil organic matter and derivative fractions by isoelectric focusing. Recent Developments in Chromatography and Electrophoresis, 10, 75-81 (1980).

4. Metodi normalizzati di analisi del suolo. II Commissione della Società Italiana della Scienza del Suolo,Firenze (1977).

5. M. De Nobili, F. Petrussi, B. Ceccanti, P. Sequi. Use of the isoelectric focusing technique (IEF) for the characterization of organic substances of different origin and as mean of following humification processes. Proceedings of the International Meeting "Humus et Planta VIII", Prague (1983).

TABLE I

TOTAL ORGANIC CARBON CONTENT OF SOME SEWAGE SLUDGE AND COMPOST SAMPLES AND THEIR FRACTIONS

Material	Total (% dry weight)	$Na_4P_2O_7$ extract mg/ml	% of total	% of extract not adsorbed on Polyclar AT	adsorbed on Polyclar AT
Sewage sludge:					
US1	15	2.3	15	39	61
US2	16	3.3	21	52	48
US3	15	3.7	25	41	59
Compost:					
UC1	20	2.8	14	63	37
UC3	n.a.	1.2	—	36	64
L	6.5	1.0	16	12	88
P	24	3.3	14	63	37

n.a. = not available.

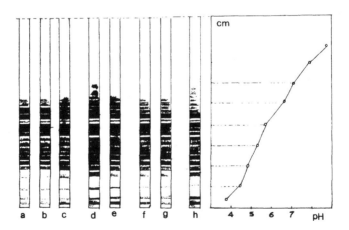

Fig.1 - IEF of 0.1M sodium pyrophosphate extracts of a),b),c) soil organic matter from A horizons of different soils, d) fresh and e) mature farm yard manure, f),g), worm composts, h) poultry manure.

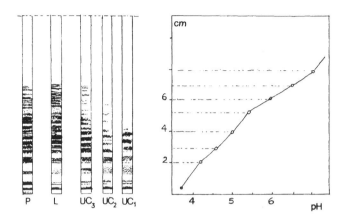

Fig.2 - IEF of 0.1M sodium pyrophosphate extracts of composts.

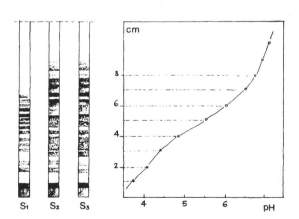

Fig.3 - IEF of 0.1M sodium pyrophosphate extracts of sewage sludges.

EVALUATION OF URBAN AND ANIMAL WASTES AS SOURCES OF PHOSPHORUS

B. POMMEL

Institut National de la Recherche Agronomique, Station d'Agronomie,
33140 Pont de la Maye, France

Summary

The organic wastes as a whole, coming from agriculture, from towns or from food industry constitute a considerable reserve of phosphorus, the fertilizing value of which needs to be determined. In order to appreciate the specific value of wastes and take into account the effects of different possible treatments, we perfected a biological test in which we measured a P uptake by plant along time, depending only on phosphorus supply, without any interaction with soil and climate.

MATERIAL AND METHODS

The plant, an hybrid ray grass (Lolium hybridum var. Sabrina) was planted in small pots containing the P source mixed with inert sand. Wastes under study were the following :
- two biological sludges, from aerated tank, anaerobically digested (d1s and d2s). They are produced in two different towns and d2s has a higher manganese and zinc content;
- the same sludge, that in addition went through a treatment with iron chloride and lime before dewatering on vacuum filter (1s);
- a chemical sludge, from flocculation and decantation after injection of iron chloride and lime; the sludge was then thickened and filtered on band (chs).
- a sludge mixed with sawdust and then composted (cos);
- a garbage compost, six months old (gc);
- a poultry manure (F);
- poultry droppings (V).

Main characteristics of these wastes are indicated in table 1.

Two insoluble mineral fertilizers were studied for comparison :
- Thomas slag (s);
- tricalcium phosphate (E).

A soluble mineral fertilizer, i.e. monocalcium phosphate, was used for reference. Wastes and fertilizers are supplied at incremental levels from 0 to 112 mg phosphorus per pot (1 kg).

Nevertheless, phosphorus is the only limiting factor for plant growth throughout the experiment. Each treatment is repeated four times.

The pots received throughout the experiment the nutrient solutions, minus P, as indicated by LEMAIRE (1977), and are placed in a growth chamber at 22°C during the day and 15°C during the night.

Water was supplied by capillarity : a nylon fabric passed through two slits in the bottom of the pot, which was placed into a similar pot that acted as a water reservoir. During the course of the experiment, demineralised water was added to the surface of the pots to a weight that represented the sand being at field moisture capacity and the lower pot containing 350 ml water.

Ray grass tops are harvested every fortnight up to depletion of the phosphorus sources that stop the growth of the plant.

RESULTS

At each cut, the cumulative phosphorus uptake, for each treatment, is a linear function of the P supplied dose, as waste or as fertilizer, with a correlation coefficient higher or equal to 0,98 (Fig. 1).

The corresponding regression lines were calculated. The ratio between slopes of regression lines for the waste or fertilizer under study on the one hand, and the monocalcium phosphate on the other hand gives the relative efficiency of the waste or the fertilizer for the relevant period of time. Thess relative efficiencies vary with time, according to the following functions that explain 98% of variations :

$$E = k \ t^C$$

where : E   represents the relative percentage efficiency,

       t   represents the time of contact between the roots and the phosphorus sources, within limits, outside of which the behaviour of the plant is disturbed. At very low values, the root-system is not sufficiently developed, and at too high values (300 days) the depletion of the substrate leads to an extreme deficiency of plant phosphorus (concentrations lower than 0,78 %.).

k and c   are coefficients peculiar to each organic waste. The first one represents the apparent efficiency after the first day; so, it is a coefficient of the intensity of P nutrition, that varies between 0 and 100. The c parameter is always positive, and expresses the growth with time of the relative P supply from wastes, as a result of the quicker depletion of soluble fertilizer, and also of a better effectiveness of root activity that allows the utilization of phosphatic forms inaccessible in the short term.

The values of k and c for each specific waste, and the calculation of E for t = 50 days are indicated in table 2. The relative efficiencies (E) are plotted against time in a logarithmic scale in figures 2 and 3.

DISCUSSION

On these sand cultivation conditions slags (s) are a good phosphorus fertilizer, the P of which is 80% to 90% as efficient as monocalcium phosphate P. This would be tested on different soils, especially calcareous soils.

Sludges composted with sawdust (cos) have a lower P content than digested sludges but a higher P availability, comparable with slags. The high microbial activity generated by the supply of an energy source probably liberates phosphorus from bonding with mineral fraction, better than root activity can do. Microbial organisms are decomposed after death and phosphorus is then present in a mineral root-available form. On the other hand, the addition of glucose to sludge creates a too short effect on microbial life to increase phosphorus availability.

Chemical sludges (chs) have a higher P content than digested sludges and show in our experimental conditions a higher P availability : they are likely to constitute an interesting source of phosphorus.

In the very short term, digested sludges (dls and d2s) are a poor phosphorus fertilizer. After 3 weeks of root-phosphorus sources contact their phosphorus is 20% in the first case, 35% in the second as efficient as m.c.p. phosphorus. This increases with time up to 60% in our experiments. Therefore, digested sludges have to be regarded as a slow release phosphatic fertilizer.

Liming of digested sludges does not affect significantly the availability of its phosphorus in the standard conditions of the experiment.

At last, garbage compost (gc) produces extremely poor intensity of phosphorus nutrition; the poultry manure shows a behaviour comparable with tricalcium phosphate, an insoluble inorganic fertilizer, again, the addition of chips of wood of the litter, constituting a slow evolving C source, had increased the phosphorus availability of poultry droppings.

CONCLUSION

The modelization in a growth chamber of P uptake by plant, along time, according to the nature of the P source, constitute a reliable test of phosphorus availability. It shows, among other results, the positive effect of composting sludge with a slow evolving carbon source on the availability of its phosphorus.

REFERENCES

LEMAIRE, F., 1977, Nouvelles observations sur l'appréciation de la fertilité des sols cultivés au moyen de l'expérimentation en petits vases de végétation. Ann. Agro 28, 425-444.

	Digested sludge d1s	Limed sludge 1s	Digested sludge d2s	Chemical sludge chs	Composted sludge cos	Garbage compost gc	Slag s	Poultry manure F	Poultry droppings V
Humidity p.100	95	79	-	80	24	42	50		
pH	6.9	9.9	-	8.1	6.4	8.1	12.6		
Ashes p.100	56.2	60.9	54.4	58.5	29.1	72	100	31	43
Carbon (Anne) p.100	27.0	24.2	29.2	-	36.1	-	tr.		
Nitrogen Kjeldahl p.100	2.68	2.33	2.79	1.66	1.92	1.03	tr.	1.1	2
Total phosphorus(P) p.100	2.12	1.55	2.1	2.73	0.51	0.4	6.68	0.75	1.3
Potassium(K) P.100	0.24	0.16	-	0.04	0.26	0.35	tr.	1	1.6
Calcium p.100	5.23	13.4	5.4	18	1.63	4.82	38.9	1.2	2.3
Magnesium p.100	0.48	0.42	0.47	0.45	0.35	0.18	1.15	0.19	0.35
Iron p.100	1.75	3.39	4.72	5.98	0.77	3.94	14.8	0.26	0.33
Copper ppm	212	216	309	112	260	2640	tr.	190	420
Manganese ppm	408	346	9250	719	532	1080	23300	180	300
Zinc ppm	2590	1890	7060	570	510	3720	tr.	130	210
Cadmium ppm	tr.	tr.	33	tr.	tr.	tr.	tr.		
Chromium ppm	336	263	74	48	1640	536	1950		
Nickel ppm	81	121	246	12	97	198	tr.		
Lead ppm	587	437	1310	79	152	720	tr.		

Table 1 Characterization of tested products

Table 2 - Recapitulation of relatives efficiencies (E = kt^c) of all
tested products

Product	k	c	r²	Period of time t in days	Efficiency % E for t = 50
Monocalcium phosphate......	100	0	automatically		100
Slag........................	57	0,12	0,98	20 - 80	91
Composted sludge...........	35	0,23	0,99	20 - 80	86
Chemical sludge............	14	0,37	0,98	20 - 80	60
Poultry manure.............	20	0,37	0,98	45 - 73	58
Tricalcium phosphate.......	12	0,37	0,98	45 - 115	51
Poultry droppings..........	13	0,32	0,98	45 - 73	45
L imed sludge	9,1	0,4	0,99	20 - 80	44
Digested sludge d1s	1,8	0,82	0,99	20 - 80	44
Digested sludge d2s........	17	0,24	0,99	20 - 300	43
Digested sludge(d1s)+glucose	1,6	0,76	0,99	45 - 115	31
Garbage compost............	781,10^{-5}	1,9	0,99	35 - 80	13

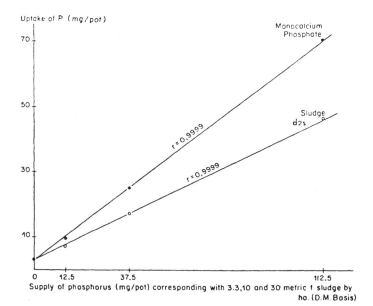

Figure 1
        Cumulative uptakes of phosphorus by the
first 15 cuttings of ray-grass, as a function of
form and dose of supply.

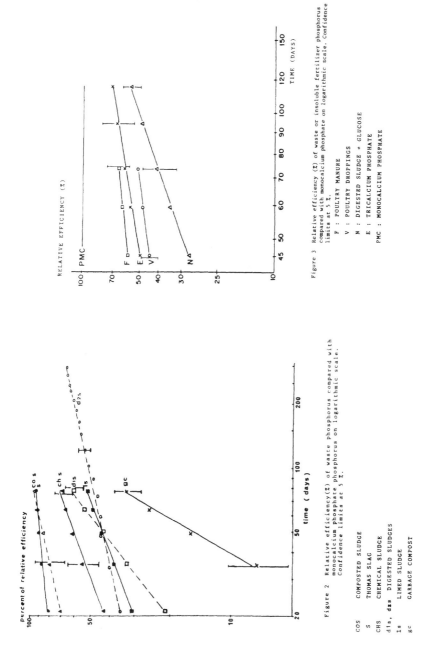

Figure 3 Relative efficiency (%) of waste or insoluble fertilizer phosphorus compared with monocalcium phosphate on logarithmic scale. Confidence limits at 5 %.

F : POULTRY MANURE
V : POULTRY DROPPINGS
N : DIGESTED SLUDGE + GLUCOSE
E : TRICALCIUM PHOSPHATE
PMC : MONOCALCIUM PHOSPHATE

Figure 2 Relative efficiency(%) of waste phosphorus compared with monocalcium phosphate phosphorus on logarithmic scale. Confidence limits at 5 %.

COS   COMPOSTED SLUDGE
S   THOMAS SLAG
CHS   CHEMICAL SLUDGE
dis, dis DIGESTED SLUDGES
ls   LIMED SLUDGE
gc   GARBAGE COMPOST

SLURRY-METER FOR ESTIMATING DRY MATTER AND NUTRIENT CONTENT OF SLURRY.

HUBERT TUNNEY
Agricultural Institute, Johnstown Castle Research Centre,
Wexford, Ireland.

Summary

In studies of the composition of animal slurry it was found that there is a good positive correlation between dry matter and plant nutrient content. In further studies to obtain a simple field test to estimate dry matter we found that there was a highly significant straight line relationship between specific gravity and dry matter of animal slurries. Based on this relationship a patented hydrometer, calibrated in percent dry matter, was developed. This can be used under field conditions to get a rapid estimate of the dry matter of slurry and the corresponding nitrogen, phosphorus and potassium content for the slurry can be read from an accompanying table. To date several hundred of these Slurry-Meters have been manufactured and distributed. The practical experience to date with the slurry-meter indicates that it is a simple to use method for estimating dry matter and fertiliser value of slurry. It is particularly valuable for estimating the quantity of slurry to apply to meet crop needs and for deciding on a price when buying or selling slurry. The feed back from people who have used the slurry meter has been positive.

1. INTRODUCTION

There has been a significant amount of research work on the fertiliser value of animal manures carried out in the E.E.C. countries over the past 10 years. A considerable part of this work has been co-ordinated and partly funded by the European Commission. Most of this work has been devoted to liquid manures or slurry from intensive animal production. Liquid manure is a valuable source of plant nutrients, however, it can be a serious source of pollution if it is not managed and used with care.

The only economical method of slurry disposal is to spread it on farm land. It is often regarded as being inferior to chemical fertilisers which are relatively inexpensive and easier and more pleasant to work with. Apart from being rather unpleasant to work with the major problems with using slurry as compared with chemical fertilisers are a) large volume with relatively low nutrient concentration, b) wide variation in nutrient composition and c) difficulty in achieveing an accurate even spread.

This paper deals with a method of overcoming the problem of variation in composition and describes a simple method that can be used under field conditions to estimate the dry matter and fertiliser value of slurry. It is based on a hydrometer, calibrated in percent dry matter, that was developed at my Institute some years ago, it is patented (1) and is generally referred to as a Slurry-Meter.

2. BACKGROUND

As a prelimainary step in studying the fertiliser value of slurry, samples were collected from 58 cattle and pig farms to determine their plant nutrient content (2). The results showed a wide variation in composition between farms, and that there was a good positive correlation between dry matter and plant nutrient content of the slurries (3).

Figure 1 illustrates the variation in dry matter composition of slurry between farms and the relationship between dry matter and nitrogen content for both cattle and pig slurry. Figure 2 illustrates the relationship between dry matter and phosphorus content of the slurry samples. There was also a significant correlation between dry matter and potassium content. These relationships are based on nutrient content in slurry samples collected on a number of Irish farms. It would be interesting to investigate how these relationships hold for other countries. Work by Steffens and Vetter (4) in Germany also indicates a wide variation in the plant nutrient content of slurry. In addition this work shows a good relationship between the dry matter and nitrogen content of slurry and it is in good agreement with results shown here in Figure 1.

Based on the results in figures 1 and 2 it was clear that the dry matter content of slurry would be a good guide to its fertiliser value. We therefore started to investigate the possibility of developing a simple field test for estimating the dry matter of slurry. We tried a number of methods based on colorimitry, conductivity and specific gravity. We found a very good straight line relationship between the specific gravity and dry matter of the slurries.

3. PRINCIPLE OF SLURRY-METER

The slurry-meter is based on the straight line relationship that exists between specific gravity and dry matter of slurry. This relationship is illustrated in figure 3. Dry matter was determined by drying 100g of slurry sample in an oven at 85°C for 24 hours.

Initially the specific gravity of the slurry samples was measured using a glass hydrometer with a specific gravity range of 1.000 to 1.050. Specific gravity was also determined volumetrically by weighing 500mls of slurry in a volumetric flask. There was good agreement between the two methods of measurement and the results were not significantly different.

This relationship between specific gravity and dry matter was first established for pig slurry and later it was found that cattle slurry showed the same relationship. Further tests with slurries prepared with flour, potatoes and other materials suggest that there is a basic relationship between specific gravity and dry matters for all slurries of plant origin.

A number of glass hydrometers calibrated in % dry matter were made in the laboratory and used in conjunction with tables based on data similar to that shown in figure 1 and 2 to estimate the fertiliser value of slurry

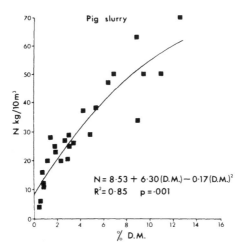

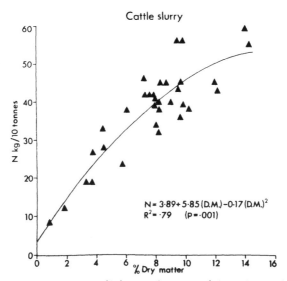

Fig. 1 :-   Relationship between dry matter and nitrogen in pig and
cattle slurry

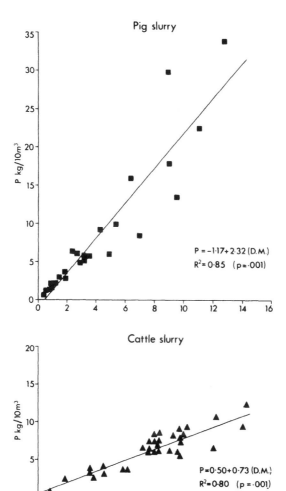

Fig. 2   Relationship between dry matter and phosphorus in
pig and cattle slurry

under field conditions.

Figure 4 shows the slurry meter being used to estimate the dry matter of two pig slurry samples.

The results in figure 3 suggest that the slurry meter estimate of dry matter should be accurate to within one unit of dry matter. For example if a reading of 5% is obtained using the Slurry-Meter the result by oven drying should be between 4% and 6%.

4. HOW TO USE A SLURRY METER

Collecting a slurry sample for estimation by Slurry-Meter requires the same care in obtaining a representative sample as for chemical analyses in the laboratory. The instructions that accompany the Slurry Meter are shown in the Appendix.

The main points are that the slurry should be well mixed immediately before the meter reading is taken to ensure that all the solid matter is in suspension.

It was found that at higher dry matter the slurry was often too viscous for accurate measurement of specific gravity with the hydrometer. The upper dry matter limit for accurate measurement varies with the type of slurry. For example it is often possible to measure dry matter up to 8% with pig slurry but cattle slurry over 6% is normally too viscous for direct measurement. Some sewage sludges with as low as 3% dry matter may be too viscous for direct measurement.

This problem can be easily overcome by mixing the slurry with an equal volume of water and then doubling the dry matter reading obtained with the Slurry-Meter. For higher dry matter slurries or for very viscous materials further dilutions may be necessary.

5. PRACTICAL EXPERIENCE

The slurry meters are made by a Hydrometer Manufacturing Company, G. H. Zeal Ltd., Lombard Road, Merton, London, England.

The Slurry-Meter is a standard glass hydrometer calibrated from 0 to 8% dry matter at 20°C.

Several hundred units have been manufactured and distributed to date; principally to Canada, Ireland and The United Kingdom with small numbers to several other countries.

I distributed a questionnaire to a number of people who have been using the Slurry-Meter. Unfortunately I received only a few replies to date, however, these indicate that the users found the Slurry-Meter helpful, easy to use and fragile.

They are fragile and easy to break particularly under farm yard conditions and when used by people with little or no experience of fragile laboratory equipment. The Slurry-Meter is resistant to corrosion by the slurry and with adequate care can give many years of satisfactory use. Copper and nickel coated metal slurry meters have been tested but are sub-

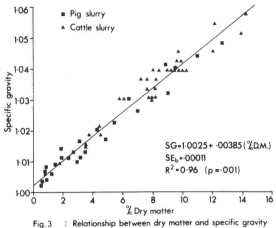

Fig. 3 : Relationship between dry matter and specific gravity
of pig slurry (25 farms) and cattle slurry (33 farms)

Fig.4 : Slurry Meter used to estimate dry matter of low (left)
and high (right) dry matter pig slurries.

ject to corrosion due to the hydrogen sulphide in the slurry.

In summary the practical experience indicates that the Slurry-Meter, despite being fragile, is a satisfactory method for obtaining a rapid estimate under field conditions of the dry matter and fertiliser value of slurry.

6. CONCLUSIONS

The Slurry-Meter has been developed by the Agricultural Institute at Johnstown Castle Research Centre. It would be interesting at this stage to have the relationships between dry matter and nutrient content studied in laboratories in other countries. The relationship between specific gravity and dry matter could also be verifed in the same studies.

The Slurry-Meter would, perhaps, be less fragile if manufactured in plastic. Manufacture in plastic would involve considerable initial cost for moulding, but the unit cost should then be lower than for glass.

7. LITERATURE

1. British Patent No. 1,543,223; 20 April 1976. A method and device for measuring the dry matter content of slurry.

2. Tunney, H. and Molloy S. M., 1975. Variations between farms in N, P, K, Mg and dry matter composition of cattle, pig and poultry manures. Ir. J. agric. Res. 14, 71-79.

3. Tunney, H., 1979. Dry matter, specific gravity and nutrient relationships of cattle and pig slurry. In: Engineering Problems with Effluents from Livestock. Ed. J.C. Hawkins. Proc. E.E.C. Seminar, Cambridge, England. Publ. C.E.C. Luxembourg. pp430-447.

4. Steffens, G. and Vetter, H., 1983. An economical use of slurry, the best way to prevent a pollution of the environment. In: Animal Waste Utilization. Proc. 4th Consultation, F.A.O. Cooperative Network, Budapest Hungary. Publ. Swedish National Board of Agriculture. pp115-143.

APPENDIX - SLURRY METER INSTRUCTIONS

1. Collect representative sample of cattle or pig slurry.

2. Put slurry in container and stir well.

3. Place slurry meter in slurry and read dry matter.

4. Read N, P and K content from Table 1.

5. Cattle slurry higher than 5% dry matter (d.m.) and pig slurry higher than 8% is normally too viscous for accurate measurement. In such cases, mix slurry with an equal volume of water, stir well and take d.m. reading with slurry meter. Double the reading obtained to get the correct d.m. of the undiluted slurry.

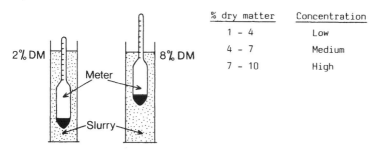

% dry matter	Concentration
1 - 4	Low
4 - 7	Medium
7 - 10	High

Table 1: Total N,P and K content of Cattle and Pig slurry in kg/10m³ (lbs/1000 gals) at various dry matter levels.

Cattle Slurry				Pig Slurry		
N	P	K	% Dry Matter	N	P	K
15	1.5	20	- 2 -	20	3	12
25	3	30	- 4 -	30	8	17
35	4.5	40	- 6 -	40	12	20
40	6	50	- 8 -	50	17	22
45	7.5	55	- 10 -	55	22	22
47	9	60	- 12 -	60	26	22

Slurry can vary from 1 to 15% dry matter. The higher the d.m. the higher the fertiliser value. Values in Table 1 are empirical data based on farm surveys of cattle (beef and dairy) fed mainly on silage supplemented with meals and pigs fed on commercial pig meal. With a different diet the figures could vary, e.g. cattle fed on meals only would produce slurry higher in P and lower in K. If a slurry sample containing mostly urine is collected the N and K values would be higher and P values lower than shown in Table 1. Ensure that the sample collected is representative of the slurry being tested.

224

DISCUSSION

Chairman: Mr J H Williams, MAFF, UK

A SUSS to M DE NOBILI

Do the individual bands in the separation relate to particular humic acid fractions?

Answer: Pure humic substances which give only one or very few bands on isoelectric focalisation are not available at present. We suggest that the isoelectric focusing pattern of soil organic matter could be used as an example of well humified and stabilised organic material. We can only compare patterns with those obtained from typical types of organic substances.

S DE HAAN to M DE NOBILI

Why was not a stronger alkali used to extract the organic acid fractions?

Answer: Sodium pyrophosphate was used for two reasons. The first was that a mild extractant was chosen to avoid possible auto-oxidation of the extracted organic fraction and the second was that we also wanted to determine enzymatic activities on the same extracts. Moreover, as far as isoelectric focalistion of well humified organic matter is concerned, if the extracts are carefully purified, differences between the NaOH and $Na_4P_2O_7$ extracts are so small that they are not relevant for this purpose.

A SUSS to B POMMEL

Did you include aluminium treated sewage sludge in your studies and how did its P availability compare with other sludges?

Answer: Aluminium salts were expensive to use for sludge treatment and none had been included amongst the sources of phosphorus used in this study.

J HALL to B POMMEL

What were your reasons for choosing a 50 day period - was this related to P availability under field conditions over one year?

All the materials you tested have variable amounts of both total and available nitrogen. How was this taken into account in the fertiliser given to your pots?

Answer: The calculations of relative efficiencies of wastes according to the indicated equations, were made, as an example, for t = 50 days. We think that this period is long enough to allow a significant uptake of phosphorus by the plant under field conditions, and sufficiently short to express an intensity of the supply of phosphorus by the source under study. This choice of time interval appears to give a good comparison between the different phosphorus sources.

To answer your second point, the amount of available nitrogen supplied by the wastes was balanced by an equivalent excess of nitrogen in nutrient solutions applied on the control treatments at the beginning of the cultures. Subsequently, the amounts of nitrogen supplied every fortnight were so high in comparison with the initial amounts that these initial differences are completely masked.

J HALL to H TUNNEY

With liquid raw sludge, we found that there was a direct relationship between increases in dry matter and increases in total nitrogen and felt sure that a hydrometer would give a good estimate of nitrogen value where the ammonia content was very low. With digested sludge, however, the ammonia content can be very high – up to 70% of the total nitrogen and, for a given sludge, the ammonia concentration tends to be independent of dry solids and only organic N increases; here the hydrometer may not be very useful. Have you any comment?

Answer: The hydrometer gives a good estimate of sludge dry matter and will give a guide to total N where it is correlated with dry matter. We found that some sewage sludges are very viscous and therefore difficult to measure with slurry meter. This would merit further studies.

E KEPPAINEN to H TUNNEY (Comment)

I believe that Dr Tunney is conducting some very relevant studies in trying out rapid methods for assessing slurry nutrient value. Inspired by these studies I have carried out some work in Finland on the same subject. I found that the correlation between dry matter content and total N and between dry matter with total P was quite good although not as good as in Dr Tunney's study. It was interesting that the correlation between dry matter and soluble N was much worse than that between dry matter and total N. This is due to variation in the solubility of nitrogen in slurry. I think we need to further investigate rapid methods for predicting the soluble N content of slurries.

L BARIDEAU to H TUNNEY

Do you think it is possible to improve the relationship between dry matter and nitrogen content by using a convenient variable transformation such as logarithmic or square root transformation. This should lessen the dispersion of the outlying points for the high values of dry matter.

Answer: There are possibilities for improving the relationship between nutrient content and dry matter. The additional values from Denmark, Finland, Canada, Germany and elsewhere should allow us more scope to try out transformations and improve the accuracy of the relationships.

K SMITH to H TUNNEY (General discussion)

In these investigations into the use of the slurry meters Dr Tunney did emphasise the difficulty of collecting a representative sample. It is a view that MAFF advisers in the UK have also held and, for this reason, we tend to make use of average nutrient contents of manure in our advice to farmers rather than attempting to sample and analyse them. We feel that we can make a satisfactory allowance for the dilution of slurries using figures such as livestock numbers, rainfall data and open yard or store area and wash water input. It is even possible with some experience to make an estimate of the dry matter content of the slurry from its visual appearance. This was possible, with experience, provided that the slurry was not very thick but still felt that there was some merit in a rapid estimation if it was reproducible.

A SUSS to S DE HAAN (General discussion)

Your results on the long term effects of sewage sludge showed mainly negative effects on sugar beet and also on potatoes. Quality parameters are definitely influenced. How is this evidence likely to affect the future of sewage sludge application in agriculture?

Answer: Maybe the negative effects were a bit over emphasized in my presentation. They had more to do with crop quality than with yield and are not specific for sewage sludge, but common for organic manures in general. The sponsors of this research project (the Water Authorities) expect that its results will contribute to an enlargement of the possibility to use sludge on arable land insofar as it will allow larger amounts to be applied at one time (say 8 instead of 2t dry matter/ha/year which has been allowed in the Netherlands up until now) although less frequently (once every 4 years instead of every year).

SUMMARY AND CONCLUSIONS

The seminar was not only devoted to the longer term beneficial effects of sewage sludges but also to those of animal slurries. Both have much in common in general but there are differences which should not be neglected in terms of nutrient release from the organic matter and in the residual benefits to be obtained. Whilst evidence was not presented on the complementary effects of sludges and slurries it can be envisaged that they could be agronomically beneficial in some situations. The main components of both products from the point of view of utilisation are the organic matter and the nitrogen which they contain. Sewage sludges are also useful sources of phosphorus for crops in the longer term whilst animal slurries can provide adequate amounts of potash for many cropping situations. Availability of the nitrogen obviously depends on the state and composition of the organic matter.

From the papers presented one must come to the conclusion that the residual effects of the nitrogen in sludges and slurries in general are small and very variable. There are occasions, however, when climatic and soil conditions and timing of applications are favourable when the residual effects can be appreciable. These effects are much more evident in those situations where sludges are applied regularly every year over a period of 3 or 4 years when the cumulative residual effects are reflected in improved grass production for example over the mid to late season period. To obtain any large residual N benefits from single applications of sludges or slurries, they must be applied in such large amounts that could prove harmful to the environment.

In the case of phosphorus, both materials have been shown to be capable of maintaining soil phosphate status but animal slurries would appear to be better able to improve the "available" soil P values than sewage sludges and can be satisfactorily maintained for at least 2 and possibly 3 years. Providing there is a good reliable method of estimating the "available" soil phosphorus, use can be made of it to give guidance on where savings on phosphate fertiliser could be made.

Evidence was presented of improved soil physical properties such as porosity and stability but, to have any significant effects, dressings need to be fairly large and repeated at short intervals of time. This could well be important on some soils in the context of erosion prevention.

Some unexplained differences in results obtained from studies on changes during storage and mineralisation of soil organic matter indicate that there are unidentified factors which affect the likely pattern of carbon and nitrogen mineralisation following the incorporation of sewage sludge or animal waste slurries and that we are still up against the difficulty of predicting with any accuracy the likely residual effects to be obtained from these materials. In the practical on-farm situation, some rapid procedures for giving an estimate of the dry matter content of farm slurries to obtain an approximation of their N and P content could prove very useful as evidenced by the results of studies from three different countries. This at least would reduce the number of variables and go some way in improving the accuracy of predicting the agronomic value of sludges and slurries. Other factors obviously play a part such as sludge or slurry type, soil, climate and timing of application all of which affect mineralisation. Some attempts need to be made to draw up a model incorporating the different factors involved in order to try and improve our prediction of their agronomic value under different circumstances.

LIST OF PARTICIPANTS

BARIDEAU, L.
Groupe Valorisation des Boues
Faculté des Sciences de l'Etat
5800 - GEMBLOUX
BELGIUM

BESSON, M.
Eidgenössische Forschungsanstalt für
Agriculturchemie und Umwelthygiene

3097 - LIEBEFELD-BERN
SWITZERLAND

CERCIGNANI, G.
Istituto di Produzione Vegetale
Università di Udine
 - UDINE
ITALY

COPPOLA, S.
University of Naples
Istituto di Microbiologia Agraria
80055 - PORTICI
ITALY

DE HAAN, S.
Institute for Soil Fertility
P.O. Box 30003
0750 RA - HAREN (Gr.)
THE NETHERLANDS

DE LA LANDE CREMER, L.C.N.
Instituut voor Bodemvruchtbarheid
Oosterweg 92
P.O. Box 30003
9750 RA - HAREN
THE NETHELANDS

DE NOBILI, M.
Istituto di Produzione Vegetale
Università di Udine
 - UDINE
ITALY

DEMUYNCK, M.

162, Chaussée d'Haecht
1030 - BRUXELLES
BELGIUM

DESTAIN, J.-P.
Station de Chimie et de Physique
Agricoles de l'Etat
115, Chaussée de Wavre
5800 - GEMBLOUX
BELGIUM

DUMONTET, S.
Istituto di Microbiologia Agraria
Università di Napoli
80055 - PORTICI-NAPOLI
ITALY

FÜRRER, O.J.
Forschungsanstalt für Agriculturchemie
und Umwelthygiene
Schwarzenburgstr. 155
3097 - LIEBEFELD (BE)
SWITZERLAND

GUIDI, G.
Laboratoria per la Chimica del Terreno
C.N.R.
Via Corridoni, 78
56100 - PISA
ITALY

GUPTA, S.K.
Eidgenössische Forschungsanstalt
für Agrikulturchemie und Umwelthygiene
Schwarzenburgstr. 155
3097 - LIEBEFELD (BE)
SWITZERLAND

HALL, J.E.
Water Research Centre
Medmenham Laboratory
Henley Road
Medmenham
P.O. Box 16
SL7 2HD - MARLOW, BUCKS
UNITED KINGDOM

KEMPPAINEN, E.
Agricultural Research Centre
MTTK
Dept. of Agric. Chemistry
31600 - JOKIOINEN
FINLAND

KOSKELA, I.
Agricultural Research Centre
MTTK
Dept. of Agric. Chemistry

31600 - JOKIOINEN
FINLAND

LA MARCA, M.
Institute for Soil Chemistry - C.N.R.
Via Corridoni 78
56100 - PISA
ITALY

LARSEN, K.E.
Askov Experimental Station
Vejenvej 55
6600 - VEJEN
DENMARK

LEITA, L.
Istituto di Produzione Vegetale
Università di Udine
 - UDINE
ITALY

LEVI-MINZI, R.
Institute for Agricultural Chemistry
Università di Pisa
Via del Borghetto 80
56100 - PISA
ITALY

PAGLIAI, M.
Institute for Soil Chemistry - C.N.R.
Via Corridoni 78
56100 - PISA
ITALY

PARENTE, E.
Istituto di Microbiologia Agraria
Università di Napoli
80055 - PORTICI-NAPOLI
ITALY

POMMEL, B.
Institut National de la Recherche
Agronomique
149, rue de Grenelle
75341 - PARIX CEDEX 07
FRANCE

PORCEDU, E.
Istituto di Ricerca sulle Acque
C.N.R.
Via Reno 1
00198 - ROMA
ITALY

RIFFALDI, R.
Institute of Agricultural Chemistry
University of Viterbo
 - VITERBO
ITALY

ROSSI, N.
Istituto Chimica Agraria
Università Bologna
Via Giaiomo 7
40126 - BOLOGNA
ITALY

SEQUI, P.
Laboratoria per la Chimica del Terreno
C.N.R.
Via Corridoni, 78
56100 - PISA
ITALY

SMITH, K.A.
Ministry of Agriculture, Fisheries
and Food
Woodthorne, Wolverhampton
WV6 8TQ - STAFFS
UNITED KINGDOM

SPALLACI, P.
Istituto Sperimentale per lo studio
e la difesa del Suolo
Piazza d'Azeglio, 30
50121 - FIRENZE
ITALY

STADELMANN, F.X.
Forschungsanstalt für Agrikulturchemie
und Umwelthygiene
Schwarzenburgstr. 155
3097 - LIEBEFELD (BE)
SWITZERLAND

SUESS, E.
Bayerische Landesanstalt für
Bodenkultur und Pflanzenbau
Vöttinger Strasse 38
8050 - FREISING
GERMANY

SUPERSPERG, H.
Universität für Bodenkultur Wien
Institut für Wasserwirtschaft
Gregor-Menzel-Strasse 33
1180 - WIEN
AUSTRIA

TOMATI, U.
I.R.E.V. - C.N.R.
Area della Ricerca di Roma
00016 - MONTEROTONDO SCALO (ROMA)
ITALY

TUNNEY, H.
The Agricultural Institute
Johnstown Castle Research Centre
 - WEXFORD
IRELAND

VIGERUST, E.
Agricultural University
Boks 28
1432 - AS-NLH
NORWAY

VOORBURG, J.H.
Rijks Agrarische Afvalwater Dienst
Kemperbergerweg 67
6816 RM - ARNHEM
THE NETHERLANDS

WILLIAMS, J.H.
Ministry of Agriculture
Woodthorne
Wolverhampton
WV6 8TQ - STAFFS
UNITED KINGDOM

ZUCCONI, F.
Ist. Coltivazioni Arboree
Università di Napoli
Via Università 100
80055 - PORTICI-NAPOLI
ITALY

235

INDEX OF AUTHORS

Printed and bound by CPI Group (UK) Ltd, Croydon, CR0 4YY

17/10/2024

01775690-0005

.